Studies in Systems, Decision and Control

Volume 667

The series "Studies in Systems, Decision and Control" (SSDC) covers both new developments and advances, as well as the state of the art, in the various areas of broadly perceived systems, decision making and control–quickly, up to date and with a high quality. The intent is to cover the theory, applications, and perspectives on the state of the art and future developments relevant to systems, decision making, control, complex processes and related areas, as embedded in the fields of engineering, computer science, physics, economics, social and life sciences, as well as the paradigms and methodologies behind them. The series contains monographs, textbooks, lecture notes and edited volumes in systems, decision making and control spanning the areas of Cyber-Physical Systems, Autonomous Systems, Sensor Networks, Control Systems, Energy Systems, Automotive Systems, Biological Systems, Vehicular Networking and Connected Vehicles, Aerospace Systems, Automation, Manufacturing, Smart Grids, Nonlinear Systems, Power Systems, Robotics, Social Systems, Economic Systems and other. Of particular value to both the contributors and the readership are the short publication timeframe and the world-wide distribution and exposure which enable both a wide and rapid dissemination of research output.

Indexed by SCOPUS, DBLP, WTI Frankfurt eG, zbMATH, SCImago.

All books published in the series are submitted for consideration in Web of Science.

Timon Rabczuk • Cosmin Anitescu •
Somdatta Goswami • Xiaoying Zhuang •
Yizheng Wang

Scientific Machine Learning with Engineering Applications

Springer

Timon Rabczuk
Bauhaus University
Weimar, Thüringen, Germany

Cosmin Anitescu
Bauhaus University
Weimar, Germany

Somdatta Goswami
Johns Hopkins University
Baltimore, MD, USA

Xiaoying Zhuang
Leibniz University Hannover
Hannover, Germany

Yizheng Wang
Department of Engineering Mechanics
Tsinghua University
Beijing, China

ISSN 2198-4182 ISSN 2198-4190 (electronic)
Studies in Systems, Decision and Control
ISBN 978-3-032-20306-9 ISBN 978-3-032-20307-6 (eBook)
https://doi.org/10.1007/978-3-032-20307-6

This Springer imprint is published by the registered company Springer Nature Switzerland AG
The registered company address is: Gewerbestrasse 11, 6330 Cham, Switzerland

To our families....

"What we know is a drop, what we don't know is an ocean."

—Isaac Newton

Preface

The objective of this book is to provide an overview and the theoretical and computational background of machine learning methods for solving partial differential equations, their implementation, and applications in engineering and materials science. This work originates from the authors' extensive research on developing and applying machine learning techniques to simulate physical systems governed by partial differential equations. The rapid advancement of machine learning, particularly deep neural networks, has opened new avenues for solving complex engineering problems that were previously intractable or computationally prohibitive using traditional numerical methods.

The scope of this book covers the fundamentals of artificial neural networks and their application to engineering simulation problems. We provide comprehensive coverage of essential machine learning concepts, including automatic differentiation, various neural network architectures, optimization algorithms, and training strategies. The book addresses both forward and inverse problems in the context of partial differential equations. Special attention is given to physics-informed neural networks (PINNs), surrogate modeling techniques, neural operators, and transfer learning approaches. We present a detailed description of commonly encountered partial differential equations in engineering and materials science, including their governing physical laws, variational formulations, and boundary conditions. Throughout the book, we illustrate concepts with practical example codes that readers can use as templates for their own applications.

The content is written from an engineering point of view. It explains concepts and formulations and provides details on implementation through Python code examples using modern machine learning frameworks. We present classical benchmark problems and interesting numerical examples to demonstrate the capabilities and performance of the methods. The book, however, does not contain rigorous mathematical proofs concerning, for instance, the convergence properties or approximation theory of neural network-based methods. Convergence and accuracy are demonstrated numerically through specific examples. While some of the concepts discussed have been implemented in research codes and are beginning to appear in commercial software, this book focuses on the fundamental methodology and implementation rather than on the use of specific software packages.

The purpose of this book is to enable newcomers to the field of engineering simulation with machine learning to understand both the fundamentals of machine learning techniques and the underlying physical laws governing the systems they wish to model. We aim to bridge the gap between the machine learning and engineering simulation communities by providing a resource that speaks to both audiences. Traditional engineers and computational scientists will find accessible explanations of modern machine learning concepts, while data scientists and machine learning practitioners will gain insight into the physical constraints and domain knowledge that are essential for successful engineering applications.

The book is aimed at advanced undergraduate students, graduate students, and researchers who are interested in learning and implementing machine learning methods for solving partial differential equations. It is well suited for students and postdoctoral fellows starting research in scientific machine learning and who are interested in method development or the application of these techniques to challenging problems in engineering and materials science. The book will also be of interest to practitioners and researchers who want to understand state-of-the-art machine learning approaches for simulating physical systems and are considering incorporating these methods into their computational toolbox. It is assumed that readers have basic familiarity with calculus, differential equations, and numerical methods, though not necessarily with machine learning or advanced computational techniques.

We would like to thank our collaborators for important discussions about the book, and in particular to Haiyue Zhao for proofreading and help with drawing some of the figures.

Weimar, Germany — Timon Rabczuk
Weimar, Germany — Cosmin Anitescu
Baltimore, MD, USA — Somdatta Goswami
Hannover, Germany — Xiaoying Zhuang
Beijing, China — Yizheng Wang

Contents

Chapter 1
Introduction

Modeling and Simulation has become a pillar in engineering and has been used as predictive tool in countless applications and branches. Thanks to the vast amount in computational science, materials science, and correspondingly, hardware and software technology, we can tackle nowadays a much larger variety of problems. However, models in Engineering and Materials Science are therefore faced with the challenge of growing complexity. Building and solving them needs increasingly interdisciplinary skills. However, as the scale and complexity of engineering problems increasingly exceed what can be solved through manual analytical methods alone, the need to utilize efficient algorithms, simulations, and decision support systems will continue to grow. These analyses and algorithms require an array of skills within different scientific disciplines to be understood and developed. Engineers and modelers will require not only in-depth knowledge of e.g. partial differential equations governing the physical field or materials behaviour but also machine learning (ML) algorithms which promise to play an increasing role in engineering science and practice. Indeed, there is an urgent need to train engineers who are also capable of systematically developing and applying Machine Learning (ML) tools for solving complex and challenging engineering problems. Even though the application of ML technologies in fields such as economics, e-commerce, image recognition and language interpretation [1] is nowadays quite advanced, it is only superficially employed in solving engineering problems. In fact, engineering fields are still not systematically covered by Artificial Intelligence (AI) despite its enormous growth, enhanced by the availability of a large amount of data, nowadays coming massively from real time monitoring of real cases as well as from outcomes of simulations allowed by High Performance Computing (HPC).

Machine Learning (ML) methods, theory and algorithms stemmed from applied mathematics and computer science. They have been recognized as efficient tools to tackle complex and urgent problems in health care, smart mobility, environment, human interactions, etc. This key technology is thus becoming part of our day-to-day lives for the benefit of all, and so its application to engineering will improve

T. Rabczuk et al., *Scientific Machine Learning with Engineering Applications*,
Studies in Systems, Decision and Control 667,
https://doi.org/10.1007/978-3-032-20307-6_1

safety, reliability, efficiency and economical aspects of a wide range of engineering challenges. ML methods have seen a very significant increase in their application to various scientific fields, in particular for cases where patterns can be extracted from complex data sets. Many of the underlying principles, such as the basics of artificial neural networks, have been developed several decades ago in the quest for useful artificial intelligence applications. However, only recently, the advances in computer hardware technology together with theoretical developments have resulted in revolutionary changes in a variety of fields such as language and image processing, medical diagnostics and drug development, as well as autonomous driving. In most cases, these advances have been "data-driven" and require systematic collection of large data sets, some of which can be laborious to obtain. However, more recently, there have been successes in developing novel algorithms that 'learn' from pre-determined rules in a more human-like fashion. A striking example of this approach is the Alpha Zero program which has achieved state of the art performance on a variety of games such as Go, Shogi and Chess using solely self-play training [2].

Machine learning algorithms can be categorized into three classes: supervised, unsupervised and reinforcement learning. In all of these classes, a set of observations or training data sets are used. In supervised learning, the training data consists of a set of input values and a corresponding set of labeled output values. Using these data sets, the ML algorithm attempts to find the best parameters of a prediction function, that can accurately represent the correspondence between the inputs and the outputs. For unsupervised learning, the output of interest is not known beforehand and the objective is to identify patterns in the given data, such as in the case of principal component analysis. Semi-supervised learning can be viewed as an intermediate between supervised and unsupervised learning, where only a subset of the input data has corresponding labeled output values. Reinforcement learning, by contrast, involves an agent that learns a behavior policy through trial-and-error interaction with an environment, optimizing for a cumulative reward signal rather than relying on a pre-labeled dataset.

Machine learning (ML) approaches for applications of engineering interest can also be roughly divided into these three categories. In supervised learning methods, a common goal is to build efficient surrogate models where the input sets are obtained in an 'offline' stage by standard (e.g. finite element) simulations or experimental data. However, the data necessary for these approaches may be expensive to obtain and the resulting models may not fully capture the physical phenomena being studied. By contrast, physics-informed models, which correspond more closely to unsupervised learning, are built from the governing equations and produce solutions without labeled training data. Hybrid approaches, analogous to semi-supervised learning, integrate experimental or synthetic datasets while also satisfying governing physical equations, which have shown some promise e.g. in modelling cardiovascular flows [3]. Nevertheless, the application of machine learning methods in 'modeling and simulation' is still in the early stages compared to other sciences. As an example, in medicine, ML-based diagnostics methods are close in accuracy with those of established physicians [4]. Machine learning has also obtained promising results in materials science [5]. Reinforcement learning has also found engineering applications, particularly in control systems, robotic motion planning, and design optimization, where agents learn policies by interacting with simulated environments and optimizing for performance-based reward signals.

As indicated in the previous paragraph, machine learning approaches in engineering can be used in data-driven contexts, as PDE (partial differential equation) solvers, or a combination of the two. Many problems in engineering can be cast in form of PDEs and a large amount of computational methods have been developed to solve them. One distinguishing feature with regard to PDE approximations is that the smoothness of the solution is directly influenced by the smoothness of the activation functions. In general, higher-order derivatives can be computed easily by automated differentiation techniques built into many machine learning software frameworks such as TensorFlow or PyTorch. Another interesting feature is that the burden of meshing can be greatly reduced since the domain discretization can be (at least in part) obtained as part of the optimization procedure on the network parameters. Therefore, neural networks can be potentially well-suited for dealing with phenomena modeled by higher-order PDEs, such as theory of Kirchhoff-Love plates and shells [6], gradient elasticity or flexo-electricity [7], to name a few. Moreover, the characteristics of ANNs make them particularly useful for highly dimensional problems, where they have been shown in some instances to overcome to a certain extent the curse of dimensionality [8].

We note that there are several different types of neural networks. The one shown in Fig. 1.1 is perhaps the simplest and also known as the feed-forward fully-connected type, where all neurons in adjacent layers are connected and the computation proceeds in one direction. Other classes of neural networks include the convolutional neural networks which are most suited for image-based or other

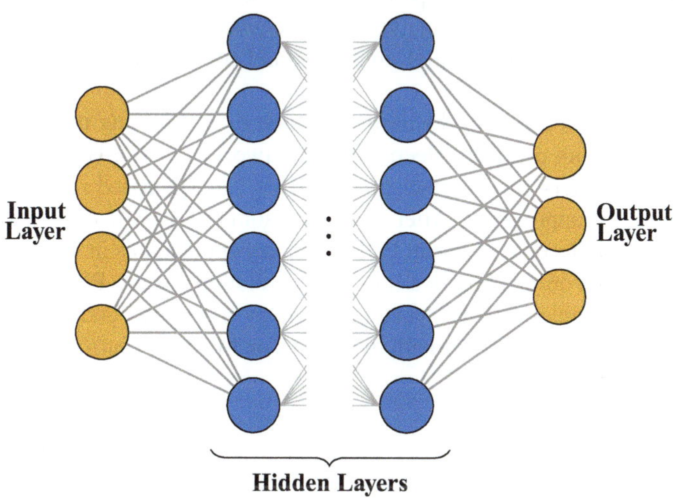

Fig. 1.1 Fully connected neural network

inputs that can be structured in a grid-like fashion. In this type of ANNs, the layers are connected in a hierarchical fashion, which allows different levels of features to be recognized at each layer. Other types of networks exist, such as recurrent ANNs, where additional time-dependent connections can be made between the neural layers enabling more efficient simulations of dynamical systems. In most cases, a goal or loss function must be defined based on a quantity of interest of the underlying problem. The minimization of the loss function is usually carried out by gradient descent or stochastic gradient descent methods [9], however quasi-Newton methods can be employed as well.

For solving a differential equation using ANNs, a loss function must be designed which determines how well the given neural network representation satisfies the governing equations and boundary conditions. For this purpose, either collocation [10] or energy minimization [6] approaches can be used. In the collocation approach, the residual of the strong form of the governing equations at suitably chosen collocation points is minimized (e.g. in a mean squared sense). Alternatively, many problems can be cast in a variational energy minimization form, which requires the evaluation of integrals over the computational domain and its boundaries. Once a suitable loss function is defined, together with a set of collocation or integration points, the weights and biases which represent the solution for a given network type can be determined, usually by using gradient-descent type algorithms. We note that, in general, the optimization problem is non-convex; therefore, stochastic gradient descent methods, which can better escape local minima and saddle points, are typically employed. The idea is to only use a random subset of gradients to determine the descent direction; however, statistical (non-deterministic) methods are needed to investigate the convergence of the minimization procedure. Quasi-Newton methods or a combination of both can be employed as well. More generally, machine learning can be seen as closely related to the differential programming paradigm [11], which aims to employ automatic differentiation methods for propagating the gradients of quantities of interest leading to efficient solvers for differential equations and optimization problems (among others).

The focus of this book will be on the solution of PDEs using machine learning approaches, or more precisely Deep Neural Networks (DNN). However, we have included two chapters on two other timely topics, i.e. ML based surrogate models and optimization approaches. After providing some basic ML concepts, we will review a set of engineering relevant PDEs with key applications in solid mechanics. Chapter 4 focuses on machine learning based solutions of the PDEs presented in Chap. 3. All examples will be 'accompanied' by associated codes with detailed explanations. Chapters 5 and 6 finally will focus on ML based surrogate models and optimization methods.

References

1. Jordan, M.I., Mitchell, T.M.: Machine learning: Trends, perspectives, and prospects. Science **349**(6245), 255–260 (2015). 10.1126/science.aaa8415
2. Silver, D., Hubert, T., Schrittwieser, J., Antonoglou, I., Lai, M., Guez, A., et al.: A general reinforcement learning algorithm that masters chess, shogi, and Go through self-play. Science **362**(6419), 1140–1144 (2018). 10.1126/science.aar6404
3. Kissas, G., Yang, Y., Hwuang, E., Witschey, W.R., Detre, J.A., Perdikaris, P.: Machine learning in cardiovascular flows modeling: Predicting arterial blood pressure from non-invasive 4D flow MRI data using physics-informed neural networks. Comput. Methods Appl. Mech. Eng. **358**, 112623 (2020). 10.1016/j.cma.2019.112623
4. Deo, R.C.: Machine learning in medicine. Circulation **132**(20), 1920–1930 (2015). 10.1161/CIRCULATIONAHA.120.050583
5. Schmidt, J., Marques, M.R., Botti, S., Marques, M.A.: Recent advances and applications of machine learning in solid-state materials science. NPJ Comput. Mater. **5**(1), 83 (2019). 10.1038/s41524-019-0221-0
6. Samaniego, E., Anitescu, C., Goswami, S., Nguyen-Thanh, V.M., Guo, H., Hamdia, K., et al.: An energy approach to the solution of partial differential equations in computational mechanics via machine learning: Concepts, implementation and applications. Comput. Methods Appl. Mech. Eng. **362**, 112790 (2020). 10.1016/j.cma.2019.112790
7. Hamdia, K.M., Ghasemi, H., Bazi, Y., AlHichri, H., Alajlan, N., Rabczuk, T.: A novel deep learning based method for the computational material design of flexoelectric nanostructures with topology optimization. Finite Elem. Anal. Des. **165**, 21–30 (2019). 10.1016/j.finel.2019.07.001
8. Grohs, P., Hornung, F., Jentzen, A., von Wurstemberger, P.: A proof that artificial neural networks overcome the curse of dimensionality in the numerical approximation of Black-Scholes partial differential equations. Preprint. arXiv:1809.02362. 10.48550/arXiv.1809.02362
9. Bottou, L.: Large-scale machine learning with stochastic gradient descent. In: Proceedings of COMPSTAT'2010: 19th International Conference on Computational StatisticsParis France, August 22–27, 2010 Keynote, Invited and Contributed Papers, pp. 177–186. Springer (2010). 10.1007/978-3-7908-2604-3_16
10. Lagaris, I.E., Likas, A., Fotiadis, D.I.: Artificial neural networks for solving ordinary and partial differential equations. IEEE Trans. Neural Netw. **9**(5), 987–1000 (1998). 10.1109/72.712178
11. Yang, Y., Aziz Bhouri, M., Perdikaris, P.: Bayesian differential programming for robust systems identification under uncertainty. Proc. R. Soc. A **476**(2243), 20200290 (2020). 10.1098/rspa.2020.0290

Chapter 2
Machine Learning Concepts

2.1 Automatic Differentiation

For solving partial differential equations, computing derivatives is a ubiquitous task as differentiating is required for the implementation of the governing equations as well as for (gradient-based) optimization and sensitivity analysis. In traditional programming, derivatives are implemented by direct (analytical) differentiation. However, as the complexity of the problem grows, so does the effort in taking the derivatives. Symbolic algebra manipulation packages can alleviate this problem to some extent by performing symbolic differentiation, however even these systems have limitations, particularly when the expressions to be differentiated contain conditional statements. Moreover, the implementation of symbolically-derived derivatives in a numerical problem can be tedious and error-prone.

An alternative to analytical differentiation is to use numerical tools, based on the finite difference method to approximate the derivatives, e.g. using the formula:

$$f'(x) \approx \frac{f(x+h) - f(x)}{h}, \tag{2.1}$$

which is valid when h is "small". However, this method is prone to numerical errors which depend on the choice of h, as h either not small enough or too small will produce inaccurate results. Moreover, this method is not efficient in case f depends on many variables and we need to compute all the partial derivatives.

A better choice for computing gradients is Automatic Differentiation (AD) [1, 2], which is at the core of many machine learning algorithms. AD can be seen as a computational tool that computes gradients automatically for a wide class of expressions, including those involving loops and conditionals that would be intractable for symbolic differentiation. Additionally, computing derivatives with AD is easier to implement, reducing the risk of errors during the implementation.

T. Rabczuk et al., *Scientific Machine Learning with Engineering Applications*,
Studies in Systems, Decision and Control 667,
https://doi.org/10.1007/978-3-032-20307-6_2

The main idea of AD is to track the algebraic operations performed on the variables that need to be differentiated, and apply the chain rule by building a computational graph, finally obtaining an exact value of the derivative. There are two major modes for AD: forward mode (FM) and reverse mode (RM). For a function $f : \mathbb{R}^n \rightarrow \mathbb{R}^m$, FM requires $O(n)$ operations, while the computational cost of RM is $O(m)$. Therefore RM is much more efficient when there are only a few outputs and many input variables. Although both methods involve the use of the chain rule, they differ on how they are implemented as described next.

2.1.1 Forward Mode

Suppose that our function z can be written as the composition of α simpler functions w_i, with $i = 1, 2, ..., \alpha$, where the first b functions $w_1, w_2, ..., w_b$ return just the values of the input variables. In FM, the derivatives with respect to the independent variables are accumulated at each intermediate step along the computational graph as $\dot{w}_i = \partial w_i / \partial s$. To obtain the gradient with respect to the first design variable, we need to first seed the values of the first b derivatives $\dot{w}_1 = 1$, $\dot{w}_2 = 0$, $\dot{w}_3 = 0$,..., $\dot{w}_b = 0$. Then the subsequent derivatives are obtained by the chain rule. For example, the value of $\dot{w}_{b+1}$ is derived as:

$$\underbrace{\frac{dw_{b+1}}{ds}}_{\dot{w}_{b+1}} = \frac{dw_{b+1}}{dw_1} \underbrace{\frac{dw_1}{ds}}_{\dot{w}_1}, \tag{2.2}$$

here, only $\dot{w}_1$ is required, since zero values were seeded for $\dot{w}_2$,..., $\dot{w}_b$. This is done repeatedly until the end of the computational graph is reached. Since $w_\alpha = z$, the last gradient $\dot{w}_\alpha$ represents the derivative of the objective with respect to the first design variable, as:

$$\dot{w}_\alpha = \frac{dw_\alpha}{ds} = \frac{dz}{ds}. \tag{2.3}$$

Figure 2.1 shows the computational graph for computing the objective function z from the design variables $\boldsymbol{s}$. For FM, the gradients $\dot{w}_i$ are calculated in the same direction as for the computation of z. It can be seen that FM requires to run through the computational graph b times in order to obtain the gradients with respect to all the b design variables. This makes FM suitable for applications where the number of inputs b is less than the number of outputs γ.

One possible shortcoming of AD is related to the memory. AD needs to be implemented carefully to efficiently use the available Random Access Memory (RAM). Since the implementation of FM involves simultaneously sweeping the computational graph of the objective function and the graph for computing the

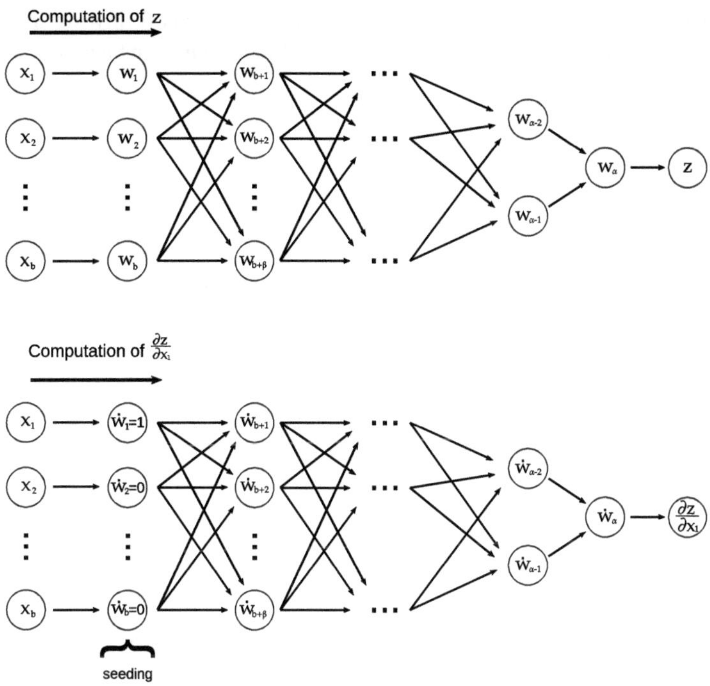

Fig. 2.1 Computational graph employed to compute the objective function and its derivatives with respect to the design variables using the forward mode of automatic differentiation

gradient for each design variable, the RAM consumption is about double that required to evaluate the objective function alone [3].

$$\dot{w}_1 = \frac{\partial w_1}{\partial w_0}\dot{w}_0, \quad \dot{w}_2 = \frac{\partial w_2}{\partial w_1}\dot{w}_1, \quad \dot{w}_3 = \frac{\partial w_3}{\partial w_2}\dot{w}_2. \tag{2.4}$$

FM can be implemented using an operator overloading approach, and making use of dual numbers to carry the gradient information. Consider a real number x and attach a component d, such that $d^2 = 0$, then the independent variables can be represented by:

$$x \mapsto x + \dot{x}d, \tag{2.5}$$

where the term $\dot{x}$ carries the value of the derivative with respect to the independent variable x. The operations performed on dual numbers are similar to those on complex numbers. To illustrate the use of dual numbers, consider the function

$$f(x_1, x_2) = x_1 \cos x_1 + x_1 \sin x_2. \tag{2.6}$$

The function can be decomposed into elementary functions following the forward propagation graph shown in Fig. 2.1. Then, each function w_i can be computed, along with its derivatives by:

$$\begin{aligned}
w_0 &= x_1 + \dot{x}_1 d, \\
w_1 &= x_2 + \dot{x}_2 d, \\
w_2 &= \cos w_0 = \cos(x_1 + \dot{x}_1 d) = \cos x_1 - \sin x_1 \dot{x}_1 d, \\
w_3 &= \sin w_1 = \sin(x_2 + \dot{x}_2 d) = \sin x_2 + \cos x_2 \dot{x}_2 d, \\
w_4 &= w_0 * w_2 = (x_1 + \dot{x}_1 d) * (\cos x_1 - \sin x_1 \dot{x}_1 d) \\
&= x_1 \cos x_1 + (\cos x_1 - x_1 \sin x_1)\dot{x}_1 d, \\
w_5 &= w_0 * w_3 = (x_1 + \dot{x}_1 d) * (\sin x_2 + \cos x_2 \dot{x}_2 d) \\
&= x_1 \sin x_2 + (\sin x_2 \dot{x}_1 + x_1 \cos x_2 \dot{x}_2) d, \\
w_6 &= w_4 + w_5 = \underbrace{[x_1 \cos x_1 + x_1 \sin x_2]}_{f(x_1, x_2)} + [\underbrace{(\cos x_1 - x_1 \sin x_1 + \sin x_2)\dot{x}_1}_{\partial f / \partial x_1} \\
&\quad + \underbrace{(x_1 \cos x_2)\dot{x}_2}_{\partial f / \partial x_2}] d.
\end{aligned} \tag{2.7}$$

As shown in the example, at each step the derivatives with respect to the variables x_i are computed.

2.1.2 Reverse Mode

RM, first described by Linnainmaa [4], exploits the chain rule for computing the derivative of a function composition as in FM. However, in this case the function f is not only a function of $g(x)$, but also of x itself. Then, in order to compute the derivative df/dx, we need to differentiate with respect to the variable x, as well as with respect to the state variable $g(x)$ using the chain rule, as:

$$\frac{df(x, g(x))}{dx} = \frac{\partial f}{\partial x} + \frac{\partial f}{\partial g} \frac{dg}{dx}. \tag{2.8}$$

Let us consider again the function to be differentiated $z(s)$ as the result of the composition of several functions, similar to FM. Then, we redefine the objective as a function of the design variable s and the state variable $w_k(s)$, such that $z(s, w_\alpha(s))$. In contrast to FM, the sensitivities dz/ds are computed from the objective function moving towards the beginning of the computational graph using Eq. (2.8), as:

$$\frac{dz}{ds} = \overset{0}{\cancel{\frac{\partial z}{\partial s}}} + \underbrace{\frac{\partial z}{\partial w_\alpha}}_{\bar{w}_\alpha = 1} \frac{dw_\alpha}{ds}, \tag{2.9}$$

in this case, the term $\partial z/\partial s$ equals zero since z does not depend explicitly on s. The seeding is performed in the last operation α, such that $\partial z/\partial w_\alpha = \bar{w}_\alpha = 1$. In RM, the intermediate derivatives $\partial z/\partial w_i$ are stored in the adjoint variables $\bar{w}_i$. The terms involving $\partial/\partial s$ are different than zero if a direct dependence appear at some point in the graph. As an example, let us recall the problem of minimization of strain energy, where $z_c = \boldsymbol{f}^T \boldsymbol{U}$. Denoting $w_\alpha = z_c$, $w_{\alpha-1} = \boldsymbol{f}^T$, and $w_{\alpha-2} = \boldsymbol{U}$, we obtain the gradient dw_α/ds as:

$$\frac{dw_\alpha(s, w_{\alpha-1}, w_{\alpha-2})}{ds} = \cancelto{0}{\frac{\partial w_\alpha}{\partial s}} + \frac{\partial w_\alpha}{\partial w_{\alpha-1}}\frac{dw_{\alpha-1}}{ds} + \frac{\partial w_\alpha}{\partial w_{\alpha-2}}\frac{dw_{\alpha-2}}{ds}, \tag{2.10}$$

Substituting into Eq. (2.9):

$$\begin{aligned} \frac{dz_c}{ds} &= \bar{w}_\alpha \left(\boldsymbol{U} \frac{d\hat{w}_{\alpha-1}}{ds} + \boldsymbol{f}^T \frac{d\hat{w}_{\alpha-2}}{ds} \right) \\ &= \underbrace{\bar{w}_\alpha \boldsymbol{U}}_{\bar{w}_{\alpha-1}} \frac{d\hat{w}_{\alpha-1}}{ds} + \underbrace{\bar{w}_\alpha \boldsymbol{f}^T}_{\bar{w}_{\alpha-2}} \frac{d\hat{w}_{\alpha-2}}{ds} \end{aligned} \tag{2.11}$$

This procedure is repeated until the adjoints $\bar{w}_1$, $\bar{w}_2$,..., $\bar{w}_b$, which are the sensitivities, are computed. Figure 2.2 shows the reverse mode for the sensitivity

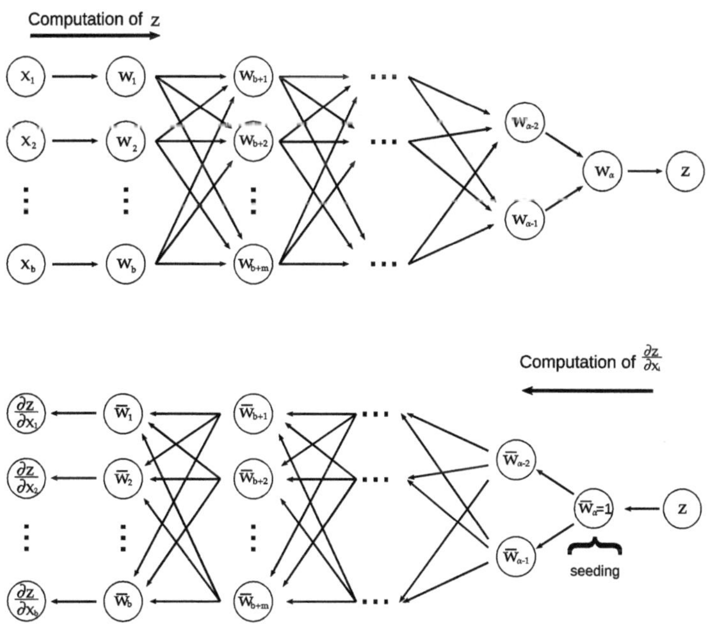

Fig. 2.2 Computational graph employed to compute the objective function and its derivatives with respect to the design variables using the reverse mode of automatic differentiation

analysis of the objective function. Contrary to FM, the sensitivities are computed in reverse order than that for computing the objective function. This is achieved by storing not only the intermediate values w_i, but also the information about the operations leading to w_i. Therefore, it is possible to obtain the gradients with respect to every design variable with a single backward pass through the computational graph (after an initial forward pass to compute and store the intermediate values). This contrasts with FM, which requires b separate forward passes for b design variables. Thus, RM is especially convenient for cases where the number of input variables is greater than the number of outputs. This is the case of shape optimization, where numerous design variables are used to compute few objective and constraint functions.

However, the implementation of RM requires more attention regarding the RAM consumption. In this case, the objective function is initially evaluated and its computational graph along with the intermediate values are stored in a persistent memory [3]. Hence, the RAM requirements of RM are higher than for FM, with the benefit of faster computational time.

By means of AD, the computation of the gradients can be notably simplified since the implementation can focus on the solution of the forward problem. In a similar fashion as in [1, 5, 6], the AD tool takes as an input the original code, where only the computational graph for obtaining the objective and constraint functions values is implemented. Next, a differentiated code, with the additional implementation of the computational graph for the gradients of such functions, is obtained. The differentiated code can be used together with an optimization algorithm for solving the structural shape optimization problem.

In order to obtain differentiated code a source code transformation approach is used. In this case, the AD toolbox has to support the specific syntax of the programming language employed in the original code [7]. This includes loops, conditionals, linear solvers, etc. Based on the primitive operations used within the commands of the host language, it is possible to implement subroutines to compute their derivatives in FM or RM, following Eqs. (2.2) and (2.8) respectively. For the case of the linear solver, this can be achieved by implementing the direct and adjoint methods for sensitivity analysis as presented in Chapter 7 in [8]. The original code is then read and transformed into a differentiated program before its execution. Once the differentiated code is obtained, it can be used in future calls without modification, unless the original code is modified as well.

2.1.3 Examples

In the following, we present several examples implemented in the Pytorch library.

2.1.3.1 Function Gradients

First, we will use automatic differentiation to compute the derivative of the function defined in (2.6). We consider the forward mode differentiation first, which is provided by the `torch.func.jvp` function. It computes the Jacobian-Vector Product (JVP), which for a function $f : \mathbb{R}^n \to \mathbb{R}^m$ is defined as:

$$\mathrm{jvp}(f)(\mathbf{v}, \mathbf{x}) = \mathbf{J}(\mathbf{x})\mathbf{v},$$

where $\mathbf{J}$ is the Jacobian matrix:

$$\mathbf{J} = \begin{bmatrix} \frac{\partial f_1}{\partial x_1} & \cdots & \frac{\partial f_1}{\partial x_n} \\ \vdots & \ddots & \vdots \\ \frac{\partial f_m}{\partial x_1} & \cdots & \frac{\partial f_m}{\partial x_n} \end{bmatrix}.$$

The jvp function takes as input three arguments, the function to be differentiated, *primals* and *tangents*. The *primals* are the variables that we differentiate with respect to (e.g. $\mathbf{x}$), and the *tangents* are the components of the vector $\mathbf{v}$ which can be chosen by the user.

A short program to illustrate the forward differentiation with `torch.func.jvp` class is shown in Listing 1. We define the function $f(x_1, x_2) = x_1 \cos x_1 + x_1 \sin(x_2)$ in lines 11–13, we initialize a grid of equally spaced points at lines 15–22 and compute the derivatives with respect to x_1 and x_2 in lines 24–34. To get the derivative with respect to each input x_k individually, we set the tangent vector v to the corresponding basis vector e_k (i.e., 1 in position k and 0 elsewhere). For example, to obtain $\frac{\partial f}{\partial x_1}$ we set $v = (1, 0)$, and for $\frac{\partial f}{\partial x_2}$ we set $v = (0, 1)$. The remainder of the program outputs the plots of the function and its derivatives, which are shown in Fig. 2.3.

The reverse mode differentiation is implemented by `torch.autograd.grad` function. It takes as input a *target* variable and a *source* variable, and computes the derivative of the target with respect to source, as shown in Listing 2. In this case, we also compute the difference between the derivatives computed by automatic (reverse mode) differentiation and the "exact" gradients, computed by symbolic algebra. When the `float32` single precision format is used, we observe a difference of the order of 10^{-7} for the derivative with respect to x_1, see Fig. 2.4a. This discrepancy is due to round-off errors in floating point algebra, since the machine accuracy limit for single precision is approximately 10^{-7} (see e.g. `numpy.finfo(numpy.float32).eps`). By changing the precision to `float64`, in lines 17-18, we obtain an improved accuracy of 10^{-16}, see Fig. 2.4b. The error for the x_2 derivative is zero, as seen in Fig. 2.4c.

This example shows that even very simple calculations are subject to numerical errors which depends on the floating point precision used. This issue is particularly relevant in scientific machine learning, since single precision, or even half-precision formats like `bfloat16` are commonly used in machine learning tasks like image classification where speed, as well as memory storage and throughput, are paramount. At the same time, most scientific computing codes use double precision format for improved accuracy and numerical stability. Therefore, some care is needed when applying machine learning algorithms to scientific computing and floating point precision is one of the important factors to keep in mind.

```python
import math
import torch
import matplotlib.pyplot as plt
from torch.func import jvp

# Function to differentiate
def f(x1, x2):
    return x1 * torch.cos(x1) + x1 * torch.sin(x2)

# Grid setup
x_min = -math.pi
x_max =  math.pi
num_vals = 51

x = torch.linspace(x_min, x_max, num_vals)
x1, x2 = torch.meshgrid(x, x, indexing='xy')

# IMPORTANT: clone to avoid zero-stride aliasing with forward-mode AD
x1 = x1.clone()
x2 = x2.clone()

# Forward-mode JVPs
y_val, dy_dx1_val = jvp(f, (x1, x2), (torch.ones_like(x1), torch.zeros_like(x2)))
_,     dy_dx2_val = jvp(f, (x1, x2), (torch.zeros_like(x1), torch.ones_like(x2)))

# 3D plotting helper
def plot3d(x, y, z, xlabel, ylabel, zlabel, rotate_z_label=False):
    fig = plt.figure()
    ax = fig.add_subplot(projection='3d')
    ax.plot_surface(x.detach().numpy(), y.detach().numpy(), z.detach().numpy())
    ax.set_xlabel(xlabel)
    ax.set_ylabel(ylabel)
    if rotate_z_label:
        ax.zaxis.set_rotate_label(True)
    ax.set_zlabel(zlabel)

# Plots
plot3d(x1, x2, y_val, '$x_1$', '$x_2$', '$y$')
plot3d(x1, x2, dy_dx1_val, '$x_1$', '$x_2$', r'$\dfrac{\partial y}{\partial x_1}$')
plot3d(x1, x2, dy_dx2_val, '$x_1$', '$x_2$', r'$\dfrac{\partial y}{\partial x_2}$')

plt.show()
```

Listing 1: Forward differentiation example

2.1.3.2 Newton's Method for Root Finding

We will use the automatic differentiation capabilities of Pytorch for a canonical application of derivatives, namely a function root finder based on Newton's method. In 1D, the statement of the problem is as follows: Given a function $f : \mathbb{R} \to \mathbb{R}$, find x such that $f(x) = 0$. Newton's method is an iterative approach, where we start with an initial guess x_0 and approximate the function by its tangent at the point $(x_0, f(x_0))$. The next guess is then the point where the tangent line intersects the x-axis. After some algebra, the method can be written as:

$$x_{i+1} = x_i - \frac{f(x_i)}{f'(x_i)} \tag{2.12}$$

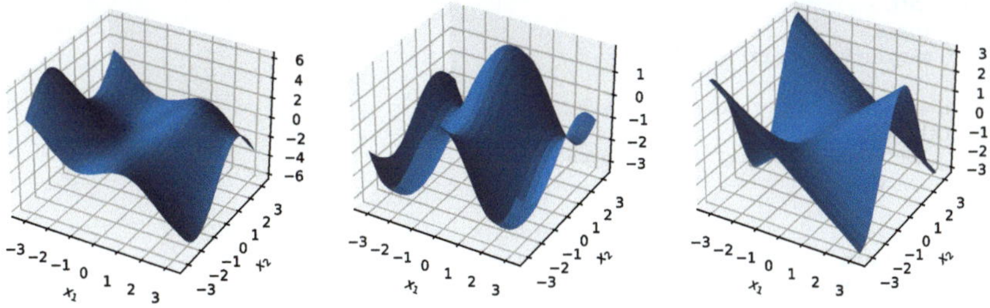

(a) $f(x_1, x_2)$ (b) $\frac{\partial f}{\partial x_1}$ (c) $\frac{\partial f}{\partial x_2}$

Fig. 2.3 Plots of the function and partial derivatives computed with forward differentiation. (**a**) $f(x_1, x_2)$. (**b**) $\frac{\partial f}{\partial x_1}$. (**c**) $\frac{\partial f}{\partial x_2}$

We take advantage of Pytorch's reverse differentiation to automatically compute the derivatives of $f(x)$. As a concrete example, we use the function

$$f(x) = x^3 - 2x - 5, \tag{2.13}$$

which was considered by Isaac Newton in his "Method of fluxions" [9]. We use as initial guess $x_0 = 0$, and set the numerical precision in to `float64`. A sample code for solving this example is given in Listing 3, and a plot of the function and the root is shown in Fig. 2.5. Since the initial guess $x_0 = 0$ is quite far away from the actual root at $x = 2.094551...$, the method needs 20 iterations during which the solution varies between -4.6 and 3.6.

We note that, in general, there is a one-to-one correspondence between common Python or Numpy expressions and the ones implemented in Pytorch, the latter of which can unlock additional performance benefits such as GPU parallelization.

2.1.3.3 Gradient-Descent and Newton Optimization

An application of automatic differentiation which is particularly relevant for machine learning is the gradient descent method used to find the minima of smooth functions. As an example, we will consider finding the minimum value of the Rosenbrock function, a standard benchmark function in optimization, which is defined as:

$$f(x_1, x_2) = (1 - x_1)^2 + 100(x_2 - x_1^2)^2. \tag{2.14}$$

Finding the minima of this function, which is at $(x_1, x_2) = (1, 1)$ is quite challenging in particular for gradient-based methods because of a wide and curved "valley" which is centered around the curve $x_2 = x_1^2$.

```python
import math
import torch
import matplotlib.pyplot as plt

# Define the function to differentiate
def f(x1, x2):
    y = x1 * torch.cos(x1) + x1 * torch.sin(x2)
    return y

# Set the limits to -pi to pi for x1 and x2 and number of plot points
x1_min = x2_min = -math.pi
x1_max = x2_max = math.pi
num_vals = 51
x1 = torch.linspace(x1_min, x1_max, num_vals)
x2 = torch.linspace(x2_min, x2_max, num_vals)
x1, x2 = torch.meshgrid(x1, x2, indexing='ij')

# Set the values of x1 and x2 to require gradients and compute derivatives
x1_val = x1.clone().requires_grad_(True)
x2_val = x2.clone().requires_grad_(True)
y_val = f(x1_val, x2_val)

# Compute gradients - use retain_graph=True for the first gradient
dy_dx1_val = torch.autograd.grad(y_val, x1_val, grad_outputs=torch.ones_like(y_val),
                                  retain_graph=True)[0]
dy_dx2_val = torch.autograd.grad(y_val, x2_val, grad_outputs=torch.ones_like(y_val),
                                  retain_graph=False)[0]

# Utility function for 3D plots
def plot3d(x, y, z, xlabel, ylabel, zlabel, rotate_z_label=False):
    fig = plt.figure()
    ax = fig.add_subplot(111, projection='3d')
    ax.plot_surface(x, y, z)
    ax.set_xlabel(xlabel)
    ax.set_ylabel(ylabel)
    ax.zaxis.set_rotate_label(rotate_z_label)
    ax.set_zlabel(zlabel)

# Plot the function to differentiate
plot3d(x1_val.detach().numpy(), x2_val.detach().numpy(), y_val.detach().numpy(),
       '$x_1$', '$x_2$', '$y$')

# Plot the error for x1-derivative
exact_dx1 = -x1_val * torch.sin(x1_val) + torch.cos(x1_val) + torch.sin(x2_val)
exact_dx2 = x1_val * torch.cos(x2_val)

plot3d(x1_val.detach().numpy(), x2_val.detach().numpy(),
       (exact_dx1 - dy_dx1_val).detach().numpy(),
       '$x_1$', '$x_2$', '     $\\dfrac{\\partial y}{\\partial x_1}$ Error')

# Plot the error for x2-derivative
plot3d(x1_val.detach().numpy(), x2_val.detach().numpy(),
       (exact_dx2 - dy_dx2_val).detach().numpy(),
       '$x_1$', '$x_2$', '          $\\dfrac{\\partial y}{\\partial x_2}$ Error')
plt.show()
```

Listing 2: Reverse differentiation example

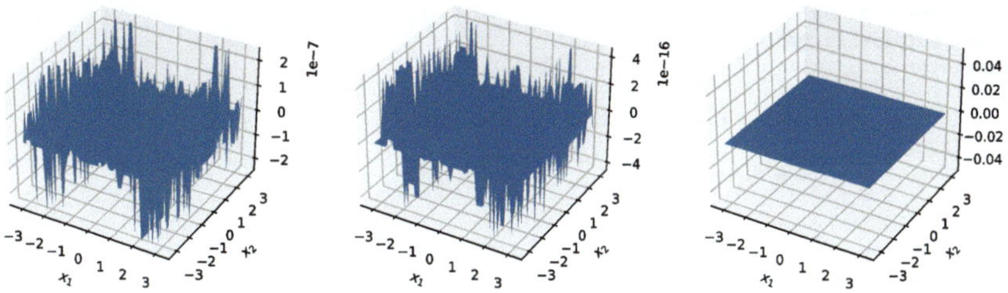

(a) Float32 error for $\frac{\partial f}{\partial x_1}$ (b) Float64 error for $\frac{\partial f}{\partial x_1}$ (c) Error for $\frac{\partial f}{\partial x_2}$

Fig. 2.4 Plots of difference between reverse mode and symbolic differentiation. (**a**) Float32 error for $\frac{\partial f}{\partial x_1}$. (**b**) Float64 error for $\frac{\partial f}{\partial x_1}$. (**c**) Error for $\frac{\partial f}{\partial x_2}$

The gradient descent method updates a current guess $\mathbf{x}_i$ by the formula:

$$\mathbf{x}_{i+1} = \mathbf{x}_i - \eta \nabla_{\mathbf{x}} f(\mathbf{x}i), \tag{2.15}$$

where η is a small number which is related to the step-size or "learning rate" of the gradient descent.

The choice of η is very important for the convergence of the gradient descent. If η is too large, then the update to $\mathbf{x}_{i+1}$ can "overshoot", leading to oscillations. In some cases, if η is too large, the next update can be in a very steep part of the function resulting in progressively larger steps in the subsequent iterations, until an overflow and eventually `NaN` (not a number) values occur. On the other hand if η is too small, the gradient descent method will require a very large number of iterations until the minimum value is found.

A listing of a simple gradient-descent minimization code using Pytorch is shown in Listing 4. The value of η is specified in line 12, and the initial guess is set as $\mathbf{x} = (-1, 2)$. Using a "large" value, like $\eta = 10^{-2}$, causes the value of $\mathbf{x}$ to blow up, eventually leading to `NaN`s. For $\eta = 2 \cdot 10^{-3}$, the method approaches the true minima at (1, 1) within 10 iterations, but then oscillates around this value and does not converge within the prescribed tolerance. For $\eta = 10^{-3}$, the output of the gradient descent takes the "long path" around the curve $x_2 = x_1^2$ and eventually converges after more than 38,000 iterations, as shown in Fig. 2.6a and b.

We note that in general the step size η does not need to be fixed between the iterations. A great deal of research has been dedicated to finding algorithms for setting η adaptively during the optimization in a way that makes the method more efficient and also more robust and capable of escaping local minima. Some of these methods will be discussed in subsequent chapters.

Another improvement over gradient descent methods are optimization algorithms using the second derivative methods, the simplest of which is Newton's method. The

```python
import numpy as np
import matplotlib.pyplot as plt
import torch

# Definition for the function for which we want to find roots (PyTorch ops)
def f(x):
    return x**3 - 2*x - 5

# Newton's method using PyTorch autograd
def newton_root_find(f, x0, tol=1e-10, max_iter=100):
    x = torch.tensor(x0, dtype=torch.float64, requires_grad=True)
    num_iter = 0
    converged = False

    for _ in range(max_iter):
        num_iter += 1
        # Forward pass
        y = f(x)

        # Backward pass to get dy/dx
        if x.grad is not None:
            x.grad.zero_()
        y.backward()  # dy/dx stored in x.grad
        dy_dx = x.grad.clone()
        delta = y / dy_dx

        # Update x: use .data (or with torch.no_grad()) to avoid tracking the update
        with torch.no_grad():
            x -= delta
        x.requires_grad_(True)  # keep tracking for next iteration
        print("x_val = ", x.item())
        if abs(delta.item()) < tol:
            converged = True
            break

    return x, num_iter, converged

# Main code
x_val, num_iter, converged = newton_root_find(f, 0.0)
if converged:
    print("Solution found: x = ", x_val.item())
    print("Number of iterations: ", num_iter)
else:
    print("Newton's method failed to converge")

# Plot the function and the root
f_np = lambda x: x**3 - 2*x - 5
xs = np.linspace(-3, 3, 100)
plt.plot(xs, f_np(xs))
plt.plot(xs, np.zeros_like(xs), color='black')
plt.scatter([x_val.item()], [f_np(x_val.item())])
plt.show()
```

Listing 3: Root finding with Newton's method

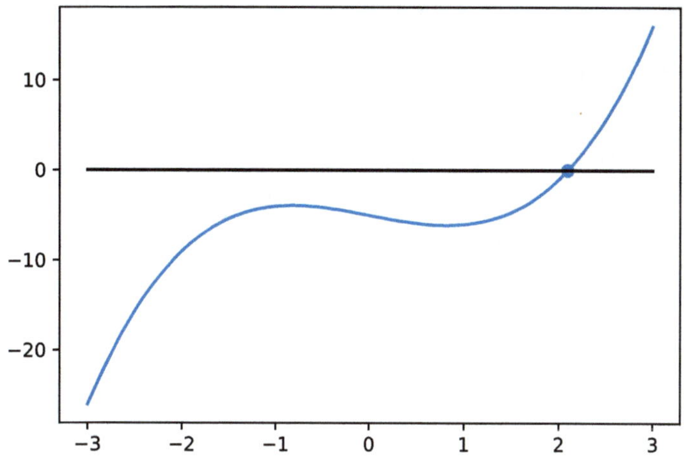

Fig. 2.5 Plot of the function $f(x) = x^3 - 2x - 5$ and its root

idea of this method is to search for the critical points, where $\nabla f(\mathbf{x}) = \mathbf{0}$ using the Hessian of the function. The method can be written as:

$$\mathbf{x}_{i+1} = \mathbf{x}_i - (\nabla^2 f(\mathbf{x}))^{-1} \nabla f(\mathbf{x}_i), \tag{2.16}$$

where $\nabla^2 f(\mathbf{x})$ is the Hessian matrix of f.

To compute the second (and higher order) derivatives in Pytorch, we can use nested `torch.autograd.grad` calls, as shown in Listing 5. Inverting the Hessian matrix can be accomplished inside Pytorch by calling the function `torch.linalg.solve`.

In contrast with gradient descent, Newton's method converges very quickly for the Rosenbrock function, within 7 iterations for this example, as shown in Fig. 2.6c. We note however that the algorithm requires computing the $n \times n$ Hessian matrix for a function $f : \mathbb{R} \to \mathbb{R}$, instead of simply the $n \times 1$ gradient vector, and is therefore much more computationally intensive when n is large. This motivates the use of quasi-Newton methods, where the Hessian is approximated in a more computationally efficient way, such as the BFGS method (suggested independently by Broyden [10, 11], Fletcher [12], Goldfarb [13] and Shanno [14] in 1970), which will be also discussed in later chapters.

2.2 Artificial Neural Networks

Artificial neural networks (ANN) can be seen as a way of storing and processing information which is loosely modeled after the biological structure of the human brain. ANNs can be "trained" to solve problems by recognizing patterns in the given

```python
import torch

# Use double precision to match the TensorFlow code
dtype = torch.float64

# Rosenbrock function
def f(x1, x2):
    return (1 - x1)**2 + 100 * (x2 - x1**2)**2

def gradient_descent_optimize(f, x_init, eta=1e-3, tol=1e-10, max_iter=50000):
    # Initialize as a single tensor with gradients enabled
    x_val = torch.tensor(x_init, dtype=dtype, requires_grad=True)
    num_iter = 0
    converged = False

    while num_iter < max_iter and not converged:
        num_iter += 1
        # Forward pass
        y_val = f(x_val[0], x_val[1])

        # Backward pass: compute gradients dy/dx
        y_val.backward()

        # x_val.grad holds the gradient; make a copy before zeroing
        grad = x_val.grad.detach().clone()

        # Update: x <- x - eta * grad
        with torch.no_grad():
            delta = eta * grad
            x_val -= delta
        print("num_iter = ", num_iter, "x_val = ", x_val.detach().numpy())

        # Convergence check
        if torch.norm(delta) < tol:
            converged = True

        # Clear gradients for next iteration
        x_val.grad.zero_()

    return x_val.detach().numpy(), num_iter, converged

# Main
if __name__ == "__main__":
    x_init = [-1.0, 2.0]
    x_val, num_iter, converged = gradient_descent_optimize(f, x_init)

    if converged:
        print("Minimum found: x = ", x_val)
        print("Number of iterations: ", num_iter)
    else:
        print("Gradient descent failed to converge within the maximum iterations")
        print("Last value found: x = ", x_val)
```

Listing 4: Finding minimum of a function with gradient descent

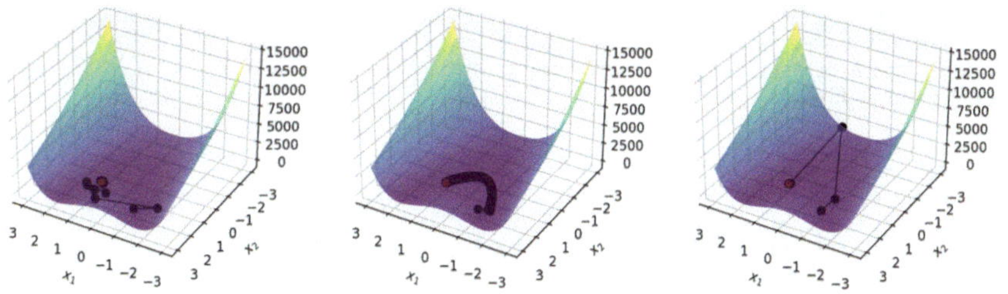

(a) Gradient descent with step size $\eta = 2 \cdot 10^{-3}$ (b) Gradient descent with step size $\eta = 10^{-3}$ (c) Gradient descent with Newton's method

Fig. 2.6 Plots of the outcome of different optimization methods. The last values obtained near the minimum (1, 1) are marked with dark red. (**a**) Gradient descent with step size $\eta = 2 \cdot 10^{-3}$. (**b**) Gradient descent with step size $\eta = 10^{-3}$. (**c**) Gradient descent with Newton's method

data, sometimes exceeding the abilities of humans or that of conventional computer programs. Several types of ANNs have been designed and successfully applied for solving a variety of problems in image and voice recognition, game playing, text processing and many other fields.

At a high level, a neural network can be seen as a function $f_{NN} : \mathbb{R}^n \to \mathbb{R}^m$, i.e. one that takes as input a vector with n components and outputs another vector with m components. The function f_{NN} depends on a set of internal parameters p, which are also called 'trainable' or 'learnable' parameters ı, which are also called "trainable" or "learnable" parameters. For a given problem, we also need to define a "loss" or "objective function" $\mathcal{L}(u_{NN}(\mathbf{x}, \mathbf{p}))$.

The building blocks of an ANN are the *neurons*, which act as computational units that aggregate weighted inputs and apply a nonlinear activation function. The learnable weights connecting the neurons encode the network's learned information and can be thought of as its memory. The neurons are organized in *layers*, with each network consisting of one *input layer*, one *output layer* and any (non-negative) number of *hidden layers*. The numbers of neurons in the input and output layers correspond to the dimensionality of the input and output data, respectively. For example, if the input is a scalar value, the input layer will have only one neuron. If the input is an image, then the input layer will have a the same number of neurons as there are pixels in the image. The number of hidden layers and the number of neurons in each hidden layer can be chosen by the user (model designer) depending on the complexity of the problem.

To illustrate the idea of function approximation in more detail, we start with a simple interpolation example. We will consider a neural network with one hidden layer with k neurons, one input and one output and we will use it to interpolate a given function $f : \mathbb{R} \to \mathbb{R}$ on an interval $x \in [a, b]$. The neural network approximator can be written as follows:

$$f_{NN}(x) = \boldsymbol{\sigma}_2(\mathbf{W}_2(\boldsymbol{\sigma}_1(\mathbf{W}_1 x + \mathbf{b}_1)) + b_2), \tag{2.17}$$

```python
import torch
# Device & dtype setup (use GPU if available)
device = torch.device("cuda" if torch.cuda.is_available() else "cpu")
dtype = torch.float64

# Rosenbrock function
def f(x1, x2):
    return (1 - x1)**2 + 100 * (x2 - x1**2)**2

def newton_optimize(f, x_init, tol=1e-10, max_iter=100):
    # x as a leaf tensor with gradients
    x = torch.tensor(x_init, dtype=dtype, device=device, requires_grad=True)
    num_iter = 0
    converged = False

    while num_iter < max_iter and not converged:
        num_iter += 1
        # Forward pass
        y = f(x[0], x[1])

        # First derivatives (gradient)
        grad = torch.autograd.grad(y, x, create_graph=True)[0]  # shape: (2,)

        # Build Hessian by differentiating each grad component
        H_rows = []
        for g_i in grad:
            # gradient of one component w.r.t x gives a Hessian row
            row_i = torch.autograd.grad(g_i, x, retain_graph=True)[0]
            H_rows.append(row_i)
        H = torch.stack(H_rows, dim=0)  # shape: (2, 2)

        # Solve H * delta = grad
        delta = torch.linalg.solve(H, grad.reshape(2, 1)).reshape(2)

        # Update x (Newton step): x <- x - delta
        with torch.no_grad():
            x -= delta
        # Keep requires_grad for next loop
        x.requires_grad_(True)
        print("num_iter = ", num_iter, "x_val = ", x.detach().cpu().numpy())

        if torch.norm(delta).item() < tol:
            converged = True
    return x.detach().cpu().numpy(), num_iter, converged

# Main
x_init = [-1.0, 2.0]
x_vals, num_iter, converged = newton_optimize(f, x_init)
if converged:
    print("Critical point found: x = ", x_vals[-1])
    print("Number of iterations: ", num_iter)
else:
    print("Newton's method failed to converge within the maximum iterations")
    print("Last value found: x = ", x_vals[-1])
```

Listing 5: Finding minimum of a function with Newton's method where $\mathbf{W}_1$, $\mathbf{W}_2$, $\mathbf{b}_1$ and $\mathbf{b}_2$ are matrices or vectors of dimensions $k \times 1$, $1 \times k$, $k \times 1$ and 1×1 respectively, and $\boldsymbol{\sigma}_1, \boldsymbol{\sigma}_2$ are "vectorized" functions of the form:

$$\boldsymbol{\sigma}_i\left(\begin{bmatrix} y_1 \\ \vdots \\ y_k \end{bmatrix}\right) = \begin{bmatrix} \sigma_i(y_1) \\ \vdots \\ \sigma_i(y_k) \end{bmatrix}. \tag{2.18}$$

Here $\sigma_i : \mathbb{R} \to \mathbb{R}$, $i = 1, 2$ are called *activation functions*, while the matrices $\mathbf{W}_i$ and vectors $\mathbf{b}_i$ are called *weights* and *biases* respectively. While the formula (2.17) may seem slightly intimidating at first, we can recognize it has a fairly simple structure: it is made up of repeated compositions of a linear transformation and an activation of the form $\boldsymbol{\sigma}_i(\mathbf{W}_i(\cdot) + \mathbf{b}_i)$, acting on vectors of dimension n_{i-1} and outputing a vector of dimension n_i, where n_{i-1} and n_i are the number of neurons in two adjacent layers.

The process of building a neural network interpolation (or more generally any approximation) can be summarized in the following steps:

1. Generate the training data in the form of input-output pairs $(x_j, f(x_j))$, $j = 1, \ldots, N$. This data may be further split into *training* and *test* subsets, where the training data is used to actively build the approximation while the test data is used to evaluate the quality of the approximation.
2. Decide on the *hyperparameters* of the neural networks, which include the number of neurons in each hidden layer, and the choice of activation functions. These parameters depend on the amount of training data and the expected complexity of the function to be approximated and in general require some experimentation until satisfactory results are obtained.
3. *Train* the neural network by optimizing the values of the *trainable* parameters $\mathbf{W}_i$ and $\mathbf{b}_i$ such that a loss function of the form

$$\mathcal{L}(x_1, \ldots, x_N; \mathbf{W}_i, \mathbf{b}_i) = \frac{1}{N}\sum_{j=1}^{N}(f(x_j) - f_{NN}(x_j))^2 \tag{2.19}$$

is minimized. This optimization usually involves some version of gradient descent where the partial derivatives $\frac{\partial \mathcal{L}}{\partial \mathbf{W}_i}$ and $\frac{\partial \mathcal{L}}{\partial \mathbf{b}_i}$ need to be computed. Since the number of parameters $\mathbf{W}_i$ and $\mathbf{b}_i$ can be quite large, it is most efficient to compute these derivatives by reverse-mode differentiation, i.e. *backpropagation*. In the case that testing data is used, the loss can be computed separately for the testing subset to evaluate the quality of the model.

A simple code for function interpolation is given in Listing 6. In this example, we consider the function $f(x) = \sin(\pi x)$ to be interpolated on the interval $x \in [-1, 1]$. We employ an ANN with 5 hidden neurons, with the activation functions $\sigma_1(x) = \tanh(\mathrm{x})$ and $\sigma_2(x) = x$. The hyperbolic tangent (*tanh*) is a commonly used activation function in ANNs, while a linear activation is usually used in the last layer to ensure that the outputs can take arbitrarily large or small values.

```python
import math
import numpy as np
import matplotlib.pyplot as plt
import torch
import torch.nn as nn

# ----- setup -----
torch.manual_seed(0)
torch.set_default_dtype(torch.float64)

# define the function to be interpolated
def f(x: torch.Tensor) -> torch.Tensor:
    return torch.sin(math.pi * x)

# define the input and output data set
xmin = -1
xmax = 1
numPts = 8
k = 5  # number of hidden neurons
x_train = torch.linspace(xmin, xmax, numPts)[:, None]    # (N,1)
y_train = f(x_train)                                     # (N,1)

# ----- build the model -----
model = nn.Sequential(nn.Linear(1, k), nn.Tanh(), nn.Linear(k, 1))
criterion = nn.MSELoss()
optimizer = torch.optim.SGD(model.parameters(), lr=0.01)

# ----- train the model -----
epochs = 20000
for epoch in range(epochs):
    optimizer.zero_grad()
    y_pred = model(x_train)
    loss = criterion(y_pred, y_train)
    loss.backward()
    optimizer.step()
    if (epoch + 1) % 100 == 0: # print loss every 100 epochs
        print(f"Epoch {epoch + 1}/{epochs} - Loss: {loss.item():.6g}")

# ----- test the model -----
numPtsTest = 10 * numPts
x_test_np = np.linspace(xmin, xmax, numPtsTest).astype(np.float64)[:, None]
x_test = torch.from_numpy(x_test_np)
with torch.no_grad():
    y_test = model(x_test)
    y_exact = f(x_test)

# ----- plots -----
plt.plot(x_test.numpy(), y_test.numpy(), label='Predicted')
plt.plot(x_test.numpy(), y_exact.numpy(), label='Exact')
plt.plot(x_train.numpy(), y_train.numpy(), '.k', label='Training Data')
plt.legend()
plt.show()
plt.plot(x_test.numpy(), y_exact.numpy() - y_test.numpy())
plt.title("Error")
plt.show()
```

Listing 6: Function interpolation with a fully connected neural network

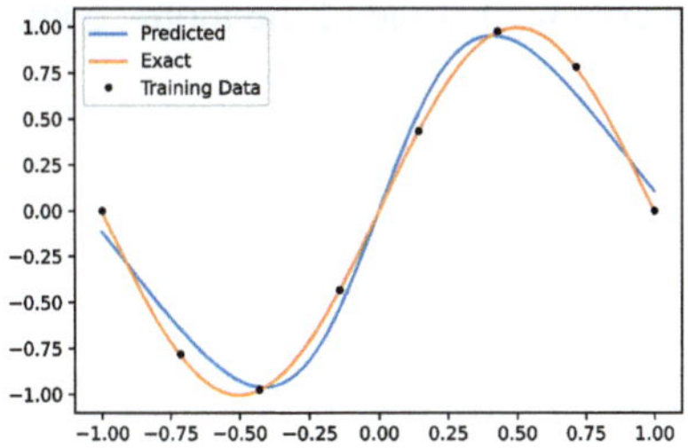

(a) Stochastic Gradient Descent (SGD) optimizer, 20000 epochs

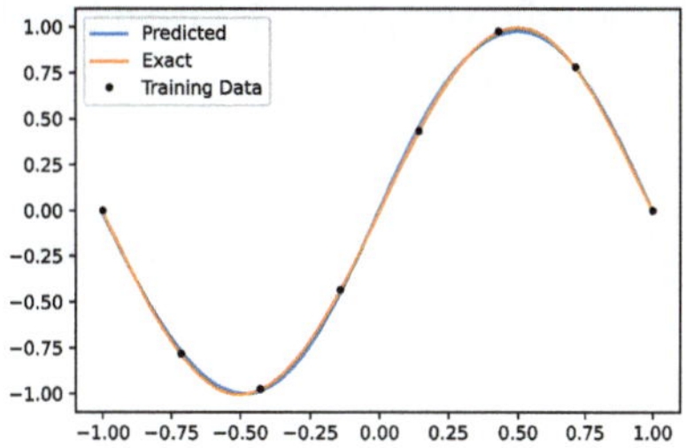

(b) Adaptive Moment Estimation (Adam) optimizer, 5000 epochs

Fig. 2.7 Predicted vs. Actual values for the 1D interpolation of the function $f(x) = \sin(\pi x)$ with SGD and Adam optimizers. (**a**) Stochastic Gradient Descent (SGD) optimizer, 20,000 epochs. (**b**) Adaptive Moment Estimation (Adam) optimizer, 5000 epochs

The setup of the network, using Pytorch is done in lines 24–26. In this setup, we also select the *stochastic gradient descent* method to optimize the parameters of the method, along with the *mean squared error* loss. For the training data, we sample the function to be interpolated at 8 equally spaced points, and store the corresponding input and output values in the `xtrain` and `ytrain` values.

The actual training of the model, which is the most time consuming part of the program, is performed in lines 30–37. For this example, we choose 20,000 training epochs, however we notice that the loss value decreases rather slowly, with a final MSE loss value of about 0.0093. When plotting the trained model, we notice a quite large disagreement between the true function to be interpolated and the predicted values (see Fig. 2.7a). The results can be improved in this case by using a different optimization algorithm, namely the Adaptive Moment Estimation (Adam) optimizer [15]. With this optimizer, the decrease of the loss is much quicker, reaching less than 10^{-4} after only 5000 training iterations or *epochs*. The only change required is changing line 26 to:

```
optimizer = torch.optim.Adam(model.parameters(), lr=5e-2)
```

The plot of the predicted function in Fig. 2.7b shows that it is also much closer and almost completely overlapping the true value.

In this example most of the "heavy-lifting" was done by a few lines of code in Pytorch neural network library (`torch.nn`). In general, e.g. for solving PDEs, we will need more complex loss functions which may include the partial derivatives of the outputs with respect to the network inputs. In this case, the Pytorch model can be extended by defining a *custom loss function*, as will be shown later in the applications. Moreover, for many problems it is advantageous to use different optimizers, for example the BFGS algorithm. In this case, a custom *training loop* can be set up to get more flexibility with regard to other optimizers and loss functions.

In the following, we will give a broad overview of some of the more commonly encountered ANNs.

2.2.1 *Multi-Layer Perceptrons*

Multi-layer Perceptron (MLP) is a type of neural network which is a generalization of the one described by (2.17) to arbitrarily many layers. In particular, we can write for a network with k layers:

$$f_{MLP}(\mathbf{x}) = \left(\boldsymbol{\sigma}_k \circ \bar{\mathbf{W}}_k \circ \boldsymbol{\sigma}_{k-1} \circ \bar{\mathbf{W}}_{k-1} \ldots \boldsymbol{\sigma}_1 \circ \bar{\mathbf{W}}_1\right)(\mathbf{x}), \tag{2.20}$$

with

$$\widetilde{\mathbf{W}}_i(\mathbf{x}) = \mathbf{W}_i\mathbf{x} + \mathbf{b}_i, \quad i = 1, \ldots, k, \tag{2.21}$$

where $\mathbf{W}_i$ and $\mathbf{b}_i$ are matrices and vectors of weights and biases respectively (the trainable parameters), and $\boldsymbol{\sigma}_i$ are the activation functions (which can be chosen by the model designer), of the type given in (2.18). A non-exhaustive list of activation functions that can be used in ANNs are given in the Pytorch documentation.[1] Some of the commonly used non-linear activation functions are as follows:

2.2.1.1 Sigmoid Activation

The formula for this activation function is:

$$\text{sigmoid}(x) = \frac{1}{1 + e^{-x}}. \tag{2.22}$$

The plot of this function is given in Fig. 2.8a. Sigmoid (also known as *logistic*) activation function) has an S-shaped plot and was originally motivated by biological neurons, where the output can be interpreted as the probability of a neuron "firing" given its inputs. Since the range of values for the outputs is between 0 and 1, this activation function is suitable for *binary classification* tasks, where we want to classify the elements of a set (inputs) into two groups.

2.2.1.2 Tanh Activation

The *hyperbolic tangent* function can be written as:

$$\tanh(x) = \frac{e^x - e^{-x}}{e^x + e^{-x}}. \tag{2.23}$$

[1] https://docs.pytorch.org/docs/stable/nn.html.

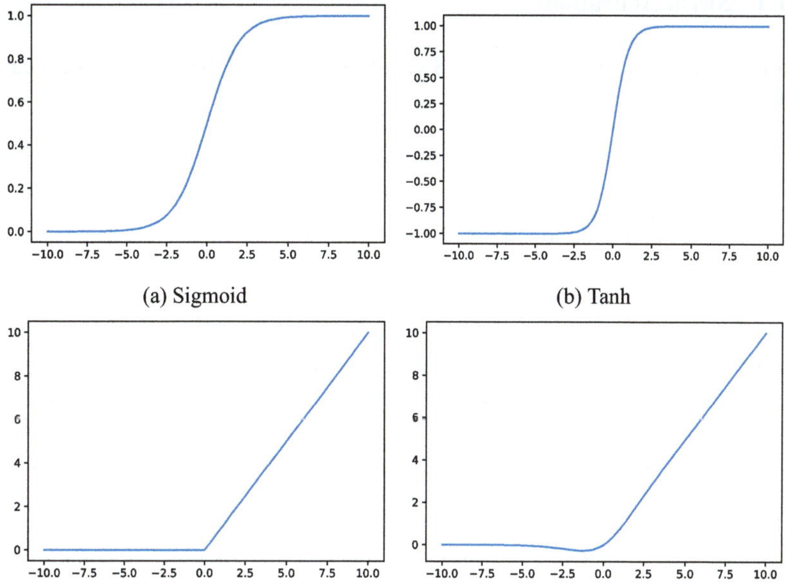

(a) Sigmoid (b) Tanh

(c) ReLU (d) Swish

Fig. 2.8 Different non-linear activation functions. (**a**) Sigmoid. (**b**) Tanh. (**c**) ReLU. (**d**) Swish

As can be seen from Fig. 2.8b, the plot of this function is also S-shaped and at a first sight, it looks similar to the one for sigmoid. However, the range of the outputs is between -1 and 1, instead of 0 and 1. This makes the tanh activation function more suitable for deep networks, without creating a positive bias as in the case of sigmoid.

2.2.1.3 ReLU Activation

The *Rectified Linear Unit (ReLU)* activation function is a piecewise linear function that can be written as:

$$\mathrm{ReLU}(x) = \max(0, x). \tag{2.24}$$

The plot of this function is shown in Fig. 2.8c. With this activation function, we can construct a piecewise linear approximation of a given function. However, a drawback is that the derivative of ReLU is not defined at $x = 0$. This can create problems, particularly in the case of differential equation approximation, where the derivatives of the output with respect to the inputs come into play.

2.2.1.4 Swish Activation

A more recent activation function is Swish [16], which in the simplest form can be written as:

$$\mathrm{swish}(x) = x \cdot \mathrm{sigmoid}(x) = \frac{x}{1 + e^{-x}}. \tag{2.25}$$

From the plot of this function in Fig. 2.8d, we can see that the overall shape is similar to that of ReLU, but the function is smooth (C^{∞}) which means that the derivatives of all orders are well-defined throughout the domain.

We note that there are several other activation functions that have been proposed, such as Exponential ReLU (ELU) [17], Mish [18], and others, some of which depend on additional user-defined or trainable parameters. While these can give a slight benefit in some cases, for constructing smooth approximations the tanh and swish activation functions are preferable. In general swish seems to be more accurate meaning it can give a better approximation for a given number of neurons and layers, while tanh is faster to train meaning the computational time per training iteration is shorter than with swish. Like with single vs. double-precision, there is a trade-off between accuracy and speed the importance of which depends in large measure on the problem at hand.

2.2.2 *Convolutional Neural Networks*

Convolutional Neural Networks (CNNs) are a category of neural networks where the input layer consists of a grid of values, which can be most commonly represented as an image or an array of values (e.g. a matrix). We note that the input can be one-dimensional or consist of more dimensions, the important feature being the existence of a spatial relation between the input values. For example, a color image in the RGB format has dimension $nx \times ny \times 3$, where nx and ny are the number of pixels in each direction and each of there $nx \times ny$ layers contain the red, green and blue intensity values of the image.

A CNN consists of three parts: convolutional layers, pooling layers and a fully-connected layer. In many neural network libraries, the convolutional layer performs a *cross-correlation* operation on a matrix input, which is defined as:

$$S(i, j) = (I * K)(i, j) = \sum_m \sum_n I(i + m, j + n) \cdot K(m, n),$$

where I is the input image, and K is a kernel of weights. The convolution operation is performed in the following steps:

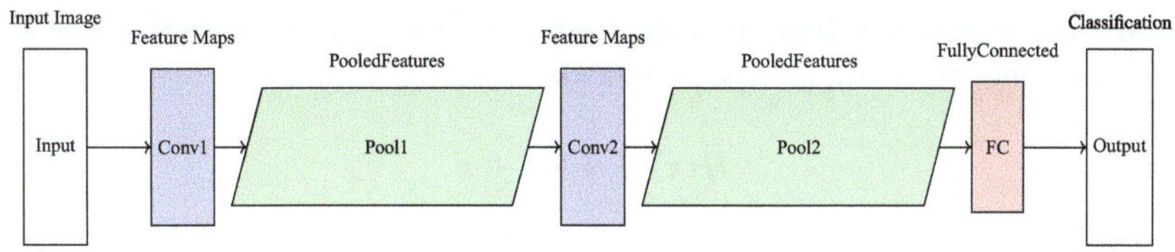

Fig. 2.9 Diagram of a Convolutional Neural Network

1. Place the kernel K on the top of the input image I at the top-left position
2. Perform element-wise multiplication between the entries in I and K and sum up all the products to get a single value
3. Place this value in the corresponding position of the output image, or *feature map*, S.
4. Slide the kernel to the right by one (or other specified *stride*) and repeat the previous steps until the entire image I is covered.

The pooling layer applies a particular operation on regions covered by the kernel. Examples are *max-pooling* which takes the maximum of each region, and *average-pooling* which uses instead the average value. Unlike, the convolutional layers, no trainable parameters are used, so the weights kernel is determined by the type of pooling operation.

Following the pooling layer, a fully-connected neural network can be used. This is useful to convert the output of the neural network to some non-image format, for example the class index in a classification task, or to output some quantity of interest based on the processed image. A diagram of a CNN is given in Fig. 2.9.

The CNN architecture can also be used as a building block for several other types of neural networks. Of particular importance are *U-nets*, where the output is also an image. The main feature of U-nets is that they use a contracting branch, where the input image is transformed into a lower-resolution array but possibly with more layers, and an expanding part, where the output of the first part is expanded back into an image of similar size as the original input. This is particularly useful in supervised scientific machine learning applications, where the output of the U-net can be used to for example to represent the solution fields [19, 20].

2.2.3 Residual Neural Networks

A Residual Neural Network (ResNet) is a type of artificial neural network designed to address the problem of vanishing or exploding gradients in deep networks, which can make training very difficult as the network depth increases. Residual Neural Networks were introduced in the context of image recognition in [21]. The idea of ResNets is to add skip connections between some of the layers of a deep neural network, by adding an earlier input tensor to the output of skipped layers. The layers containing a skip connections are called *residual blocks*. In other words, instead of learning a function $H(x)$, the residual block learns $F(x)$, where:

$$H(x) = F(x) + x.$$

Here $F(x)$ is the learned transformation (e.g. through fully-connected or convolutional layers) and x is the original input that is added directly to the output. The skip connections make it easier for the network to learn because it preserves the identity of the input if the added layers are not needed.

Residual blocks were proposed as part of the architecture of the neural network for the Deep Ritz method [22]. A more advanced type of residual connection was proposed in the context of PINNs in a neural architecture called PirateNets [23].

2.2.4 Recurrent Neural Networks

Recurrent neural networks (RNNs) are a specialized class of artificial neural networks designed to process sequential data by maintaining an internal memory of past inputs. This memory enables them to model temporal dependencies and contextual patterns in time-series data, making them particularly valuable in scientific machine learning applications where sequential or spatiotemporal relationships are critical.

RNNs operate by cycling information through a hidden state—dynamic memory bank that blends current input with historical context. At each time step, the network:

- Receives an input (e.g., a data point from a time series)
- Updates its hidden state using both the input and the previous hidden state
- Generates an output influenced by this updated state

Mathematically, an RNN can be described by the following mathematical relation:

$$h_t = \phi(W_{xh}x_t + W_{hh}h_{t-1} + b_h), \quad y_t = \psi(W_{hy}h_t + b_y), \tag{2.26}$$

where W_{xh}, W_{hh}, and W_{hy} are learnable weight matrices, b_h and b_y are biases, and ϕ, ψ are activation functions (e.g., tanh or softmax). A schematic of an RNN is shown in Fig. 2.10.

Unlike feedforward neural networks, RNNs share parameters across all time steps, allowing efficient learning of sequential patterns. Training involves backpropagation through time (BPTT), which adjusts weights by propagating errors backward across the sequence.

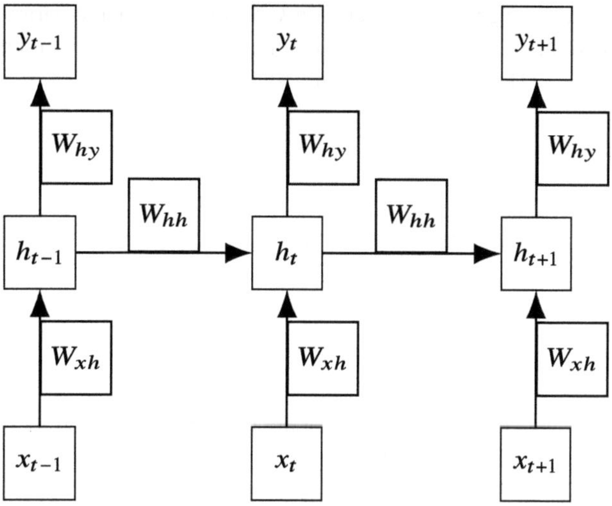

Fig. 2.10 Unrolled Recurrent Neural Network over three time steps

2.3 Optimizers

Optimization algorithms lie at the core of scientific machine learning. The trainable parameters in neural networks are determined through these algorithms. Currently, the mainstream optimization algorithms in deep learning are broadly categorized into first-order and second-order methods.

First-order methods are efficient and simple but do not account for the curvature of the loss surface near the parameters, unlike second-order methods. As a result, although second-order methods typically require fewer iterations to converge, each iteration is computationally more expensive due to the need to compute or approximate the Hessian matrix.

2.3.1 First-Order Optimization Algorithms

The most commonly used first-order optimizer is Adam (Adaptive Moment Estimation). Below we explain the principle behind Adam.

We begin with gradient descent. Given a mini-batch, the gradient is computed as:

$$\boldsymbol{g}^{(k)} = \frac{1}{M} \sum_{i=1}^{M} \left. \frac{\partial \mathcal{L}(x^{[i]}, y^{[i]}; \boldsymbol{\theta})}{\partial \boldsymbol{\theta}} \right|_{\boldsymbol{\theta}^{(k)}} \tag{2.27}$$

where $\boldsymbol{g}^{(k)}$ denotes the gradient at iteration k, $\boldsymbol{\theta}$ are the trainable parameters, and $\mathcal{L}$ is the loss function. $x^{[i]}$ and $y^{[i]}$ are the input and output of the i-th data point, and M is the batch size. The gradient descent update rule is:

$$\boldsymbol{\theta}^{(k+1)} = \boldsymbol{\theta}^{(k)} - \alpha \boldsymbol{g}^{(k)} \tag{2.28}$$

where α is the learning rate.

Next, we compute the first-order and second-order moment estimates:

$$\begin{aligned} \boldsymbol{m}^{(k)} &= \beta_1 \boldsymbol{m}^{(k-1)} + (1 - \beta_1)\boldsymbol{g}^{(k)} \\ \boldsymbol{v}^{(k)} &= \beta_2 v^{(k-1)} + (1 - \beta_2)\boldsymbol{g}^{(k)} \odot \boldsymbol{g}^{(k)} \end{aligned} \tag{2.29}$$

Here, $\boldsymbol{m}^{(k)}$ represents the exponential moving average of past gradients, and $v^{(k)}$ is the exponential moving average of the squared gradients.

The idea in Eq. (2.29) is based on exponentially weighted averages, meaning that the current estimate depends on both recent and past values, with older contributions decaying exponentially. This increases robustness and reduces oscillations in the optimization process. As shown in Fig. 2.11, optimization trajectories can vary significantly depending on the choice of algorithm and the landscape of the objective function.

Typically, β_1 and β_2 are close to 1, with default values 0.9 and 0.999, respectively. Both $\boldsymbol{m}$ and v are initialized to zero. Since the weights in Eq. (2.29) do not sum to 1 in early steps, for example:

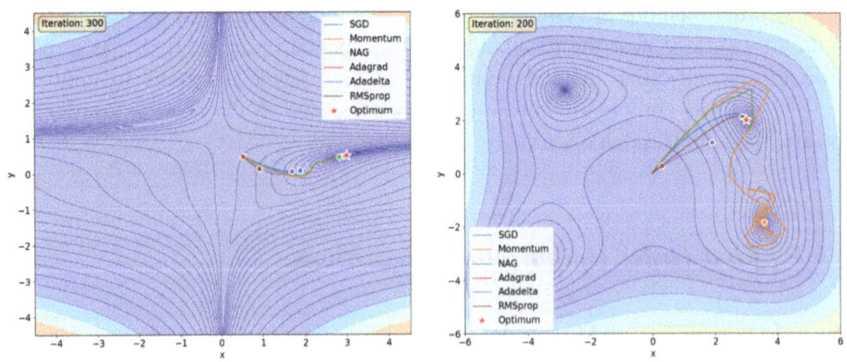

(a) Trajectories on the Beale function (b) Trajectories on the Himmelblau function

Fig. 2.11 Illustration of optimization trajectories on benchmark functions. The Beale function (**a**) has a global minimum at (3, 0.5), while the Himmelblau function (**b**) has four identical local minima. (**a**) Trajectories on the Beale function. (**b**) Trajectories on the Himmelblau function

$$\begin{aligned}
\boldsymbol{m}^{(1)} &= (1-\beta_1)\boldsymbol{g}^{(1)} \\
\boldsymbol{m}^{(2)} &= (1-\beta_1)(\beta_1\boldsymbol{g}^{(1)} + \boldsymbol{g}^{(2)}) \\
\boldsymbol{m}^{(3)} &= (1-\beta_1)(\beta_1^2\boldsymbol{g}^{(1)} + \beta_1\boldsymbol{g}^{(2)} + \boldsymbol{g}^{(3)}) \\
&\dots \\
\boldsymbol{m}^{(k)} &= (1-\beta_1)\sum_{i=0}^{k-1}\beta_1^i\boldsymbol{g}^{(k-i)}
\end{aligned} \tag{2.30}$$

we observe that the sum of the weights is:

$$\begin{aligned}
\boldsymbol{m}^{(1)} &: 1-\beta_1 \\
\boldsymbol{m}^{(2)} &: 1-\beta_1^2 \\
\boldsymbol{m}^{(3)} &: 1-\beta_1^3 \\
&\dots \\
\boldsymbol{m}^{(k)} &: 1-\beta_1^k
\end{aligned} \tag{2.31}$$

Therefore, we normalize the moments:

$$\begin{aligned}
\hat{\boldsymbol{m}}^{(k)} &= \frac{\boldsymbol{m}^{(k)}}{1-\beta_1^k} \\
\hat{\boldsymbol{v}}^{(k)} &= \frac{\boldsymbol{v}^{(k)}}{1-\beta_2^k}
\end{aligned} \tag{2.32}$$

Finally, the Adam update step is given by:

$$\boldsymbol{\theta}^{(k+1)} = \boldsymbol{\theta}^{(k)} - \alpha\frac{\hat{\boldsymbol{m}}^{(k)}}{\sqrt{\hat{\boldsymbol{v}}^{(k)}} + \delta} \tag{2.33}$$

where δ is a small constant to prevent division by zero.

Importantly, the second moment estimate v allows Adam to adaptively adjust the learning rate: when v is large, indicating large gradients, the learning rate is reduced to ensure stable updates; conversely, when v is small, the learning rate increases to accelerate convergence. In this way, Adam adaptively adjusts both the direction (via momentum $\hat{\boldsymbol{m}}^{(k)}$) and the step size (via $\hat{v}^{(k)}$), which justifies its name: **Adaptive Moment Estimation**.

2.3.2 Second-Order Optimization Algorithms

Second-order optimization algorithms are closely related to Newton's method in numerical analysis. We explain the idea from the perspective of nonlinear Newton iteration.

The goal is to minimize the loss function, i.e.,

$$\begin{aligned}\delta\mathcal{L}_{\text{batch}} &= \frac{\partial\mathcal{L}_{\text{batch}}}{\partial\boldsymbol{\theta}}\cdot\delta\boldsymbol{\theta} = 0\\ \mathcal{L}_{\text{batch}} &= \frac{1}{M}\sum_{i=1}^{M}\mathcal{L}(x^{[i]}, y^{[i]}; \boldsymbol{\theta})\end{aligned} \tag{2.34}$$

The necessary condition for Eq. (2.34) is:

$$\frac{\partial\mathcal{L}_{\text{batch}}}{\partial\boldsymbol{\theta}} = 0 \tag{2.35}$$

Equation (2.35) is a nonlinear system, which can be solved by Newton's method:

$$\begin{aligned}\boldsymbol{\theta}^{(k+1)} &= \boldsymbol{\theta}^{(k)} - \left(\boldsymbol{H}|_{\boldsymbol{\theta}^{(k)}}\right)^{-1}\left.\frac{\partial\mathcal{L}_{\text{batch}}}{\partial\boldsymbol{\theta}}\right|_{\boldsymbol{\theta}^{(k)}}\\ \boldsymbol{H}|_{\boldsymbol{\theta}^{(k)}} &= \left.\frac{\partial^2\mathcal{L}_{\text{batch}}}{\partial\boldsymbol{\theta}\,\partial\boldsymbol{\theta}}\right|_{\boldsymbol{\theta}^{(k)}}\end{aligned} \tag{2.36}$$

By introducing a learning rate α, we obtain the second-order update:

$$\boldsymbol{\theta}^{(k+1)} = \boldsymbol{\theta}^{(k)} - \alpha\left(\boldsymbol{H}|_{\boldsymbol{\theta}^{(k)}}\right)^{-1}\left.\frac{\partial\mathcal{L}_{\text{batch}}}{\partial\boldsymbol{\theta}}\right|_{\boldsymbol{\theta}^{(k)}} \tag{2.37}$$

Due to the high computational cost of inverting the Hessian, quasi-Newton methods such as BFGS (Broyden–Fletcher–Goldfarb–Shanno) are often used to approximate $\boldsymbol{H}^{-1}$. However, the memory cost of storing the full Hessian in BFGS is quadratic in the number of parameters, which is prohibitive for deep networks.

To mitigate this, the **Limited-memory BFGS (L-BFGS)** algorithm is used. L-BFGS restricts memory usage by storing only a few recent updates, making it a practical second-order optimization method for problems with a moderate number of parameters.

An example of defining the second-order optimization algorithm L-BFGS in PyTorch is illustrated in Listing 7. The parameter `max_iter` defines the maximum number of internal iterations that the optimizer will perform within a single call to `optimizer.step(closure)`. The parameter `history_size` controls the number of previous updates stored to approximate the inverse Hessian matrix, i.e. Limited-memory in L-BFGS.

```
# Define Second-order Optimizer: LBFGS
optimizer = optim.LBFGS(model.parameters(), lr=1e-3, max_iter=20, history_size=10)

def closure():
    optimizer.zero_grad()
    output = model(x)
    loss = loss_fn(output, y)
    loss.backward()
    return loss

optimizer.step(closure)
```

Listing 7: The second-order optimizer LBFGS demo

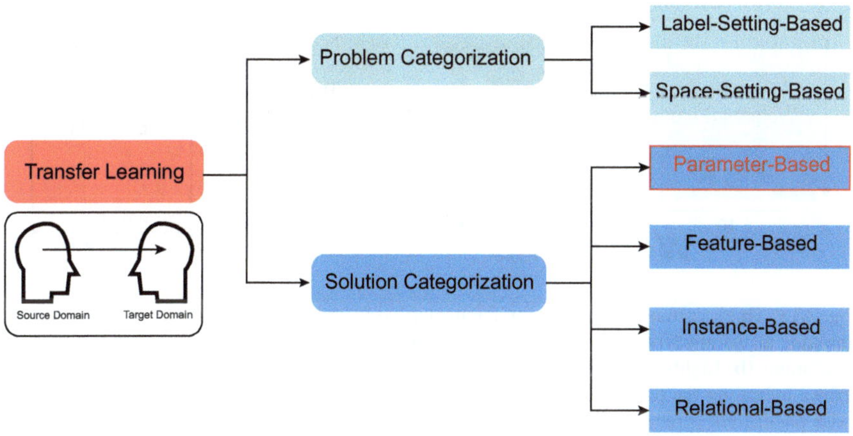

Fig. 2.12 Categories of transfer learning [24]

2.4 Transfer Learning

Transfer learning refers to fine-tuning a pre-trained model for a new related task, often requiring fewer iterations and less data for the new task [24]. Therefore, exploring transfer learning in scientific machine learning for different boundary conditions [25], material distributions [26], and geometries [27] is of great importance. The advantage of transfer learning in scientific machine learning lies in iterative algorithms, as they can inherit parameters from previous training, thereby reducing the number of iterations required for new but similar tasks [28].

Given that there are various transfer learning methods, as shown in Fig. 2.12, most of scientific machine learning focus on the classical Parameter-based transfer learning. The source domain refers to the pre-trained source problem, while the target domain refers to the target problem. The core of transfer learning lies in leveraging the parameters pre-trained on the source domain to achieve faster convergence on the target domain.

Below, two common approaches (full and lightweight fine-tuning) for Parameter-based transfer learning are introduced.

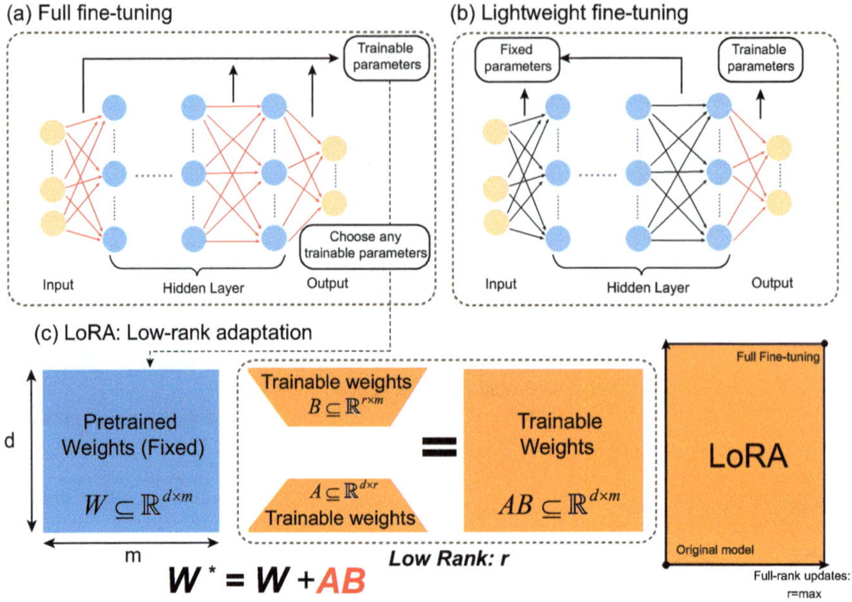

Fig. 2.13 Three common methods for Parameter-based transfer learning: (**a**) Full fine-tuning: fine-tune all parameters of the model, with the red arrows indicating the parameters that need fine-tuning. (**b**) Lightweight fine-tuning: fine-tune a subset of the model's parameters, with the red arrows indicating the parameters that need fine-tuning. (**c**) LoRA: the blue matrix $\boldsymbol{W}$ represents fixed parameters (pretrained parameters from the previous task). The yellow matrices are the trainable parameters $\boldsymbol{A}$ and $\boldsymbol{B}$, where $\boldsymbol{AB}$ is a low-rank matrix with rank r. During training on the new dataset, only $\boldsymbol{A}$ and $\boldsymbol{B}$ are adjusted. $\boldsymbol{W}^* = \boldsymbol{W} + \boldsymbol{AB}$ represents the neural network weights during testing

2.4.1 Full Fine-Tuning

The most common method for adapting pretrained models to downstream tasks is full fine-tuning as shown in Fig. 2.13a, where all model parameters are fine-tuned [29]. However, this results in a full copy of fine-tuned model parameters for each task, which becomes prohibitively expensive when the models have a large number of trainable parameters [30]. The core of full fine-tuning is to initialize the neural network parameters with those from the old task. Then, on the new task, the parameters from the old task are fine-tuned:

$$\boldsymbol{\theta}^{new} = \arg\min_{\boldsymbol{\theta}} \mathcal{L}(D^{new}; \boldsymbol{\theta}^{old}), \tag{2.38}$$

where $\boldsymbol{\theta}^{old}$ is the parameter set after the old task is completed, and $\boldsymbol{\theta}^{new}$ is the result of optimizing the loss function $\mathcal{L}$ on the new dataset D^{new}.

2.4.2 Lightweight Fine-Tuning

Lightweight fine-tuning freezes most of the pretrained parameters and only tune a smaller subset of parameters [29], as shown in Fig. 2.13b. The key question is how to decide which subset of pretrained parameters to tune.

A common method for lightweight fine-tuning is to freeze the early layers of the neural network and only train the later layers to adapt to the new task [25, 31, 32]. The motivation for this approach is that shallow layers typically extract basic features, while retraining deeper layers enables the transfer of abstract knowledge. Due to the significant reduction in trainable parameters, the computational costs are substantially reduced. Listing 8 shows an example of a lightweight fine-tuning program (only train the last layer of MLP) for the MLP architecture is [3, 100, 100, 100, 100, 2].

However, a more flexible approach, known as LoRA (Low-Rank Adaptation) [33], can be used to determine which subset of pretrained parameters to fine-tune. LoRA introduces the idea of multiplying low-rank matrices $\boldsymbol{AB}$ (where $\boldsymbol{A} \in \mathbb{R}^{d\times r}$ and $\boldsymbol{B} \in \mathbb{R}^{r\times m}$) and then adding them to the pretrained parameters $\boldsymbol{W} \in \mathbb{R}^{d\times m}$, as illustrated in Fig. 2.13c. The parameter computation for LoRA is given by:

$$\boldsymbol{W}^* = \boldsymbol{W} + \alpha \boldsymbol{AB}, \tag{2.39}$$

where r is the rank and hyperparameter in LoRA. In general, $r \ll \min(d, m)$.

It is important to note that $\boldsymbol{W}$ represents the fixed pretrained parameters from the old task (source domain). Only the product $\boldsymbol{AB}$ is trained in the target domain, with the number of trainable parameters being $r \times (d + m)$. The number of trainable parameters in $\boldsymbol{AB}$ is much smaller than the total number of parameters in $\boldsymbol{W}$, which is $d \times m$. The scaling factor α is used to control the weight between the pretrained $\boldsymbol{W}$ and the LoRA-trained parameters $\boldsymbol{AB}$, with a default value of 1. $\boldsymbol{A}$ and $\boldsymbol{B}$ are initialized using a Gaussian distribution. Other initialization methods for $\boldsymbol{A}$ and $\boldsymbol{B}$ are also possible. We employ a Gaussian distribution with a mean of 0 and a standard deviation of 0.02. Listing 9 shows an example of the LoRA program (only train the last layer of MLP) for the MLP architecture is [3, 100, 100, 100, 100, 2]. LoRA trains only the hidden layers in the middle, without training the input and output layer, i.e., LoRA only trains the layers [3, 100, 100, 100, 100, 2] on [100, 100, 100, 100].

In essence, LoRA can be seen as a more flexible version of lightweight fine-tuning. Additionally, full fine-tuning can be considered a limiting case of LoRA, when $r = \min(d, m)$. The core idea of LoRA is to decompose part of the weight matrices in the model into the product of two low-rank matrices, thereby reducing the number of parameters that need to be trained. Details about transfer learning in scientific machine learning can refer to [34].

```
import torch
import torch.nn as nn

class MultiLayerNet(nn.Module):
    def __init__(self, D_in, H, D_out):
        """
        Initializes a multi-layer neural network with 5 linear layers.
        """

def freeze_layers(model: nn.Module, train_layer_keywords: list, lr=1e-3):
    """
    Freezes all parameters in the model except for the ones whose layer
    names contain the specified keywords,
    and returns an optimizer that updates only the specified layers.

    Args:
        model (nn.Module): The PyTorch model
        train_layer_keywords (list): A list of keywords of the layer names
        that should be updated (fine-tuned).
            For example, ["linear3", "linear4"] means only the layers containing
            "linear3" or "linear4" will be updated.
        lr (float): The learning rate for the optimizer

    Returns:
        optimizer (torch.optim.Optimizer): An optimizer that updates only
        the specified layers' parameters
    """
    # 1) Iterate through all parameters and check if their layer names
    contain any of the specified keywords
    for name, param in model.named_parameters():
        #        requires_grad=True    requires_grad=False
        if any(keyword in name for keyword in train_layer_keywords):
            param.requires_grad = True
        else:
            param.requires_grad = False

    # 2) Filter the parameters that require gradients and create the optimizer
    params_to_optimize = [p for p in model.parameters() if p.requires_grad]

    # 3) Create the optimizer with the parameters that require gradients
    optimizer = torch.optim.Adam(params_to_optimize, lr=lr)

    return optimizer

# Assuming 'device' is defined elsewhere in your code (either 'cuda' or 'cpu')
model = MultiLayerNet(3, 100, 2).to(device)

# Load model weights from source domain
model.load_state_dict(torch.load('source_domain.pth'))

# Freeze all layers except linear5
optimizer = freeze_layers(model, train_layer_keywords=["linear5"], lr=0.001)
```

Listing 8: Lightweight finetuning demo: Only train the last layer of MLP

```python
import torch
import torch.nn as nn

class LoRALinear(nn.Module):
    def __init__(self, in_features, out_features, r=4, alpha=1.0, bias=True):
        """
        Low-Rank Adaptation (LoRA) version of a linear layer: W_eff = W +
        alpha * (A @ B)

        Args:
            in_features (int): Number of input features
            out_features (int): Number of output features
            r (int): Rank in LoRA
            alpha (float): LoRA scaling factor
            bias (bool): Whether to include a bias term
        """

class MultiLayerNetLoRA(nn.Module):
    def __init__(self, D_in, H, D_out, r=4, alpha=1.0):
            """
            LoRA version of a multi-layer fully connected network:
            Uses LoRALinear for every layer, and sets the same rank and alpha.
            If a layer does not want to use LoRA, you can set r to 0 for that layer,
            or directly use nn.Linear.
            """

def freeze_lora_layers(model: nn.Module, train_layer_keywords: list, lr=1e-3):
    # Iterate through all parameters
    for name, param in model.named_parameters():
        # If a keyword matches, keep requires_grad=True; otherwise,
        set requires_grad=False
        if any(keyword in name for keyword in train_layer_keywords):
            param.requires_grad = True
        else:
            param.requires_grad = False
    # Collect parameters that require gradients
    params_to_optimize = [p for p in model.parameters() if p.requires_grad]
    # Build the optimizer
    optimizer = torch.optim.Adam(params_to_optimize, lr=lr)
    return optimizer

model = MultiLayerNetLoRA(D_in=3, H=100, D_out=2, r=4, alpha=1.0).to(device)
# Initialize the parameters of the model from the source domain
model.load_state_dict(torch.load('source_domain.pth'))

# Only train the hidden layers in LoRA
optimizer = freeze_lora_layers(model, train_layer_keywords= \
["linear2.lora", "linear2.bias","linear3.lora", "linear3.bias",\
"linear4.lora", "linear4.bias"], lr=0.001)
```

Listing 9: LoRA demo

2.5 Architecture and Hyperparameter Optimization

Architecture and hyperparameters play a crucial role in determining the performance and effectiveness of a model in machine learning. A machine learning architecture refers to the structure and organization of components and processes within a machine learning system. It encompasses the design principles and flow of operations for handling data, training models, assessing performance, and generating predictions. Hyperparameters are external configuration variables that

Table 2.1 Popular machine learning framework

Framework	Institution	1st Release	Github stars (2023)
Tensorflow	Google Brain	2015	180k
Pytorch	Facebook	2016	74.2k
Keras	Individual Author	2015	60.2k
Scikit-learn	David Cournapeau	2007	57k
Caffe	Berkeley Research	2014	33.7k
MXNet	Apache Software Foundation	2015	20.7k
Microsoft Cognitive Toolkit	Microsoft	2016	17.4k
Deeplearning4j	Skymind	2014	13.3k
Theano	University Montreal	2008	9.8k
Torch	Ronan Collobert et al.	2002	8.9k
Chainer	Preferred Networks	2015	5.8k

are set prior to the training model. Hyperparameters include learning rate, batch size, number of epochs, regularization parameters, and so on.

The choice of architecture largely depends on the nature and type of a given task. Most of the time, it is an iterative and experimental process, which involves a combination of domain knowledge, intuition, and empirical testing to search for the suitable architecture. As for hyperparameters optimization, there are different strategies such as grid search, random search, Bayesian optimization, automated hyperparameter tuning, etc. Hyperparameter optimization is a resource-intensive and iterative process, which also requires a balance between exploration and exploitation in terms of cost efficiency. One needs to make informed decisions on computational budget, optimization strategies, and parallelization to efficiently discover optimal hyperparameters.

Popular artificial intelligence frameworks such as TensorFlow and PyTorch have significantly contributed to the development of machine learning models. These frameworks are designed to provide a comprehensive set of tools and abstractions, making it easier for developers to create machine learning models. The popular machine learning frameworks are listed in Table 2.1 [35].

2.5.1 *Automated Machine Learning*

When machine learning is employed to solve problems and obtain good results, practitioners need to perform data collection and preprocessing, feature engineering, model and algorithm selection. By carefully selecting the appropriate algorithm and tuning the hyperparameters, the performance of the machine learning models can be enhanced. This iterative process often involves experimentation, evaluation, and fine-tuning to strike the right balance between model complexity and generalization.

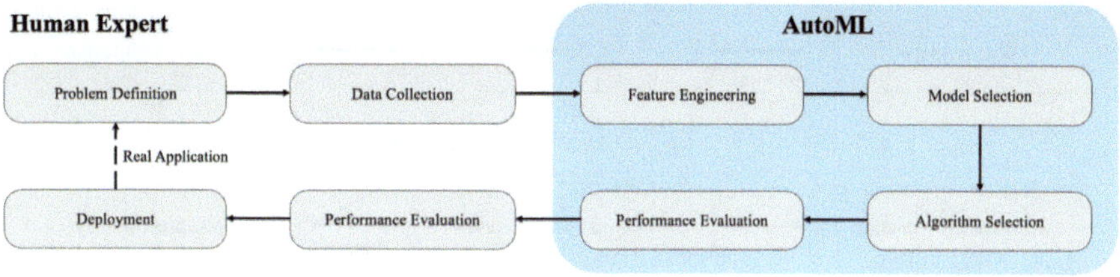

Fig. 2.14 The use of AutoML, adapted from [37]

Table 2.2 Popular AutoML tools and projects

Name	Institution	1st Release	Features
Auto-sklearn	University of Freiburg	2016	An automated machine learning toolkit and a drop-in replacement for a scikit-learn estimator
TPOT	The University of Pennsylvania	2015	Use genetic programming to optimize machine learning pipelines
H2O	H2O.ai	2012	Automated: data prep, hyperparameter tuning, random grid search and stacked ensembles in a distributed ML platform
AutoKeras	Texas A & M University	2018	Built on Keras, it provides a user-friendly API for neural architecture search
Google AutoML	Google	2018	A suite of machine learning products that enables developers to build and deploy custom ML models
MLbox	Axel de Romblay	2017	Accurate hyper-parameter optimization in high-dimensional space with support for distributed computing
AutoGluon	Amazon Web Services	2019	Support multiple machine learning frameworks

The design of neural architecture often depends on the prior knowledge, experience, and insights.

Automated machine learning (AutoML) is the process of automating the end-to-end process of applying machine learning to real-world problems. The goal of AutoML is to simplify and accelerate the machine learning workflow, enabling users to build high-performing machine learning models with minimal manual intervention, as shown in Fig. 2.14. The use of AutoML can be traced back to AutoWEKA [36], which automatically selects the learning algorithm and set its hyperparameters. Some popular AutoML tools and related projects are listed in Table 2.2 [35].

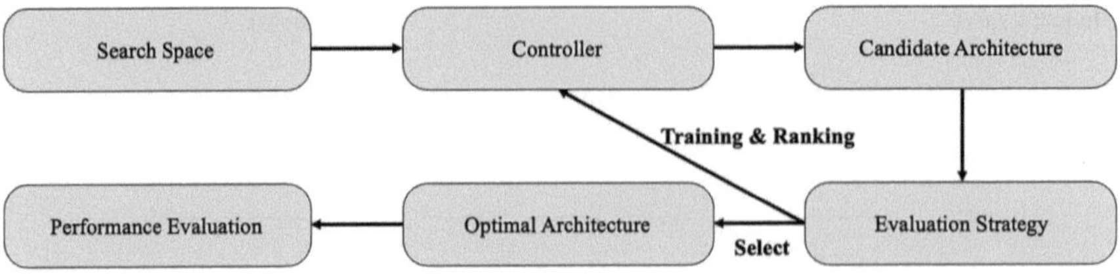

Fig. 2.15 The general framework of NAS, adapted from [39]

2.5.1.1 Neural Architecture Search (NAS)

Neural architecture search is a subfield of AutoML, which focuses on automating the design of artificial neural networks. Neural architecture search (NAS) aims to automatically discover network architectures that match or surpass those designed by human experts, and has in several cases yielded novel structural motifs that outperform hand-crafted designs. The idea behind NAS is to select the optimal neural network architecture from a predefined architecture space. NAS can effectively reduce the use and implementation cost of neural networks. Compared with the traditional manual processes, NAS ensures higher efficiency and better performance.

NAS involves three main aspects that can be categorized as search space, search strategy, and performance estimation strategy (Fig. 2.15).

- Search space: The search space defines all possible neural network architecture. It includes choices such as the number and the type of the layers, their connectivity, and hyperparameters. Incorporating prior knowledge for a task is a key strategy to reduce the size of the search space. However, it does indeed introduce the risk of human bias. It is important to find a balance between leveraging prior knowledge and allowing the NAS algorithm enough flexibility to explore novel architectures.
- Search strategy: The search strategy determines how to explore the search space, which is often exponentially large or even unbounded [38]. Typical search strategies include random search, neuro-evolutionary methods, Bayesian optimization, and reinforcement learning.
- Performance estimation strategy: Performance estimation strategy refers to evaluating the performance of each architecture. However, full training on each neural network requires high computation time and limits the number of architectures to be explored. To reduce the cost of the performance estimations, several methods can be used, including proxy task performance, low-fidelity performance estimation, weight inheritance, weight sharing, learning-curve extrapolation, and network morphism.

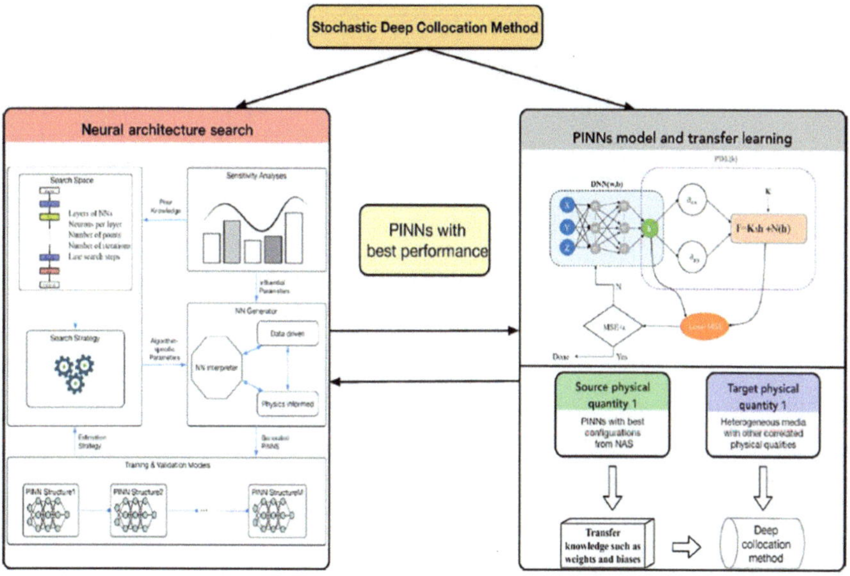

Fig. 2.16 The stochastic deep collocation method with PINN and NAS submodules [44]

2.5.1.2 Examples of Models Based on NAS

NAS has been utilized to create various models with optimized architectures in different tasks, object detection, image classification, and so on. The work of training RNNs with reinforcement learning for architecture space search plays a crucial role in popularizing NAS as a research area. AmoebaNet-A is proposed with regularized evolution for image classifier architecture search [40]. DARTS is applied to address the scalability challenge of architecture search by formulating the task in a differentiable manner [41]. A progressive NAS method of learning the CNN structure is proposed for image classification [42]. ENAS is proposed for automatic model design via computational graph construction and parameters sharing [43].

NAS is also widely used to search the neural architecture for solving PDEs, A modified NAS based physics-informed neural networks (PINNs) is presented for stochastic analysis in heterogeneous porous material [44], as shown in Fig. 2.16. The model is fitted to the deep collocation method with sensitivity analysis and transfer learning (Fig. 2.17). The approach is also used for melting heat transfer analysis [45]. Moreover, NAS guided PINN method has been validated to efficiently solve different PDEs including Poisson, Burgers, and Advection equations [46].

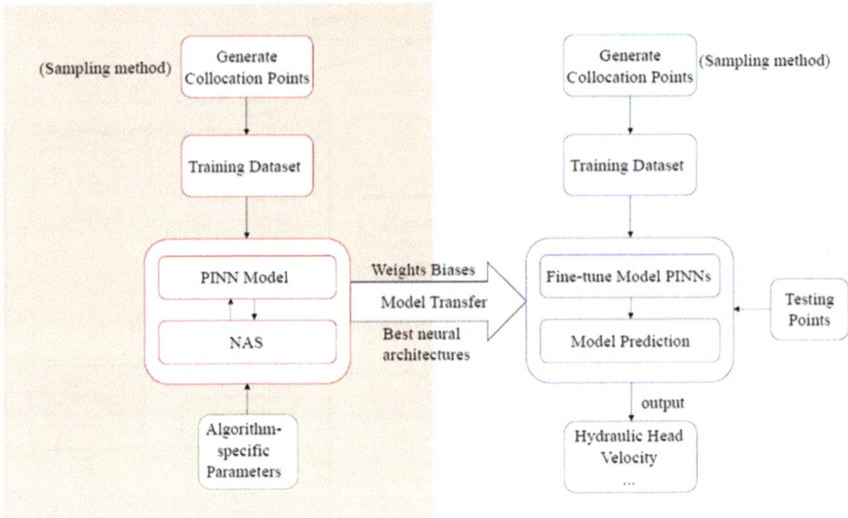

Fig. 2.17 The transfer learning schematic [44]

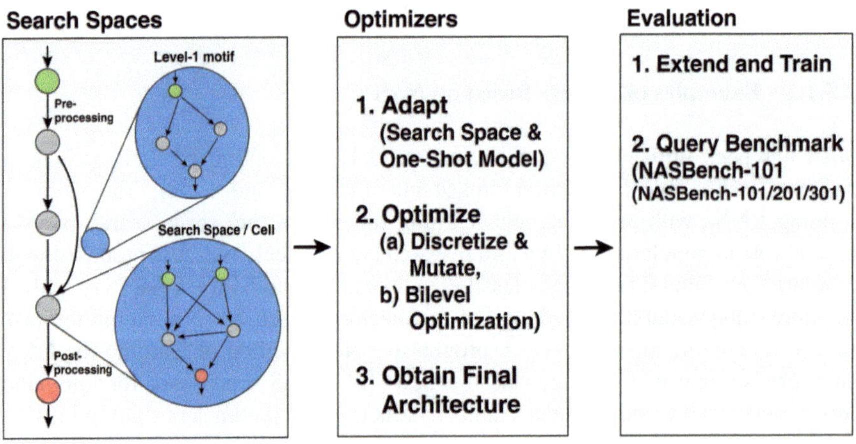

Fig. 2.18 The NASLib library, adapted from [47]

2.5.1.3 Open-Source NAS Libraries

Neural architecture search libraries are designed to facilitate research in the field of automatic neural network architecture design (Fig. 2.18). Typical NAS libraries are listed in Table 2.3.

Table 2.3 Open-source NAS libraries

Library	Institution	Platform	Features
NASLib	AutoML Freiburg group	Pytorch	A modular and flexible framework
Auto-Pytorch	Groups Freiburg and Hannover	Pytorch	Support tabular and time series data
NasBench	Google Research	Tensorflow	A NAS Dataset and Benchmark
ENAS	Google Brain	Tensorflow	Efficient NAS via parameter sharing

2.5.2 Emerging New Models and Frameworks

Artificial intelligence is evolving rapidly, and research trends include meta-learning, explainable artificial intelligence, multimodal models, generative artificial intelligence, and so on. Many new models and frameworks in computational science and engineering have emerged recently. Physics-enhanced deep-surrogate approach for complex physics and engineering applications gains a lot of attention. A combination of a low-fidelity, explainable physics simulator, and a neural network generator is proposed for solving partial differential equations [48]. There are also libraries developed for physical problems based on physics-informed machine learning approaches, such as Nvidia PhysicsNeMo [49], DeepXDE [50], IDRLnet [51], NeuroDiffEq [52], and so on.

Neural operators refer to the use of neural networks to learn nonlinear mappings between function spaces in mathematical and scientific applications. The deep operator network (DeepONet) [53] and Fourier Neural Operator (FNO) [54] are two typical neural operators. DeepONet (Fig. 2.19) consists of a DNN for encoding the discrete input function space and another DNN for encoding the domain of the output functions [53]. DeepONet has been used for predicting crack path [55], multiple input functions [56], time-dependent problems [57], and so on. FNO is based on parameterizing the integral kernel in the Fourier space [58]. It has been used for various problems such as multiphase flow [59], heterogeneous material modeling [60], the weather forecasting [61]. A comprehensive comparison of DeepONet and FNO is given in [54].

Generative artificial intelligence models, such as Google's Gemini and the ongoing OpenAI Q* project, are reshaping research priorities and applications in a variety of fields [62]. GPT has become one of the most discussed breakthroughs in the latest AI technology, which allows users to define a specific character. This feature can open up the possibility of using ChatGPT for various applications in engineering and natural sciences. A ChatGPT-based assistant framework is proposed to handle diverse problems in computational mechanics [63]. In addition, a new numerical method is developed for solving elliptic-type PDEs by adapting the Q-learning algorithm in reinforcement learning [64].

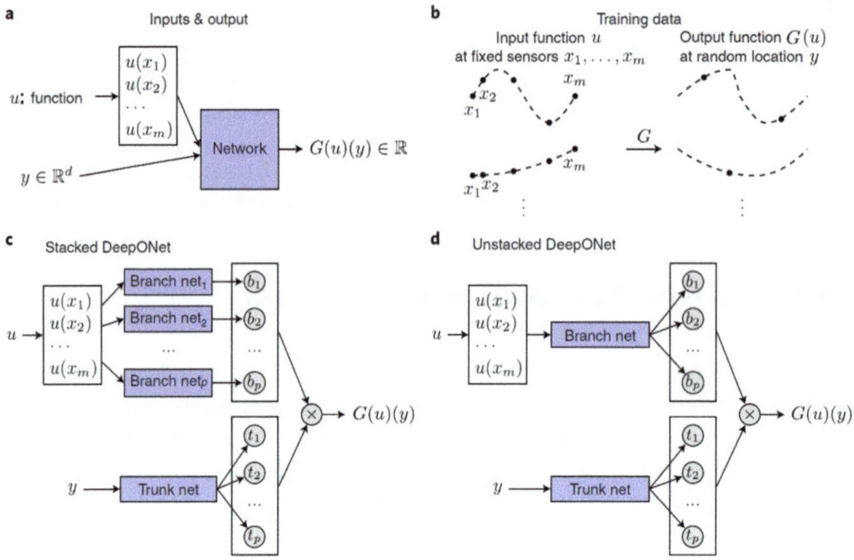

Fig. 2.19 The architectures of DeepONets [53]

2.6 Computer Implementation

In this section, we introduce the overall code framework for scientific machine learning with engineering applications, as illustrated in Fig. 2.20. The implementation of physics-informed neural networks for solving engineering problems follows a systematic workflow that can be decomposed into six main steps:

Step 1: Problem Definition The first step involves defining the problem geometry, material properties, governing equations, and boundary and initial conditions. This includes specifying the spatial domain Ω and its boundary $\partial\Omega$, identifying the relevant partial differential equations (PDEs)—which may be elliptic, parabolic, or hyperbolic depending on the physics—and prescribing boundary conditions (e.g., Dirichlet, Neumann, or Robin) and initial conditions for time-dependent problems. Material properties such as conductivity, elasticity moduli, or diffusion coefficients must also be defined at this stage.

Step 2: Loss Function Formulation Next, we construct the objective (loss) function that will guide the neural network training. The total loss typically comprises multiple components: a residual loss (or energy loss) that enforces the governing PDEs in the interior domain, a boundary loss that penalizes violations of boundary conditions, an initial condition loss for time-dependent problems, and optionally a data loss term when experimental or simulation data are available. The loss function can be formulated in either strong form (PDE residual) or weak

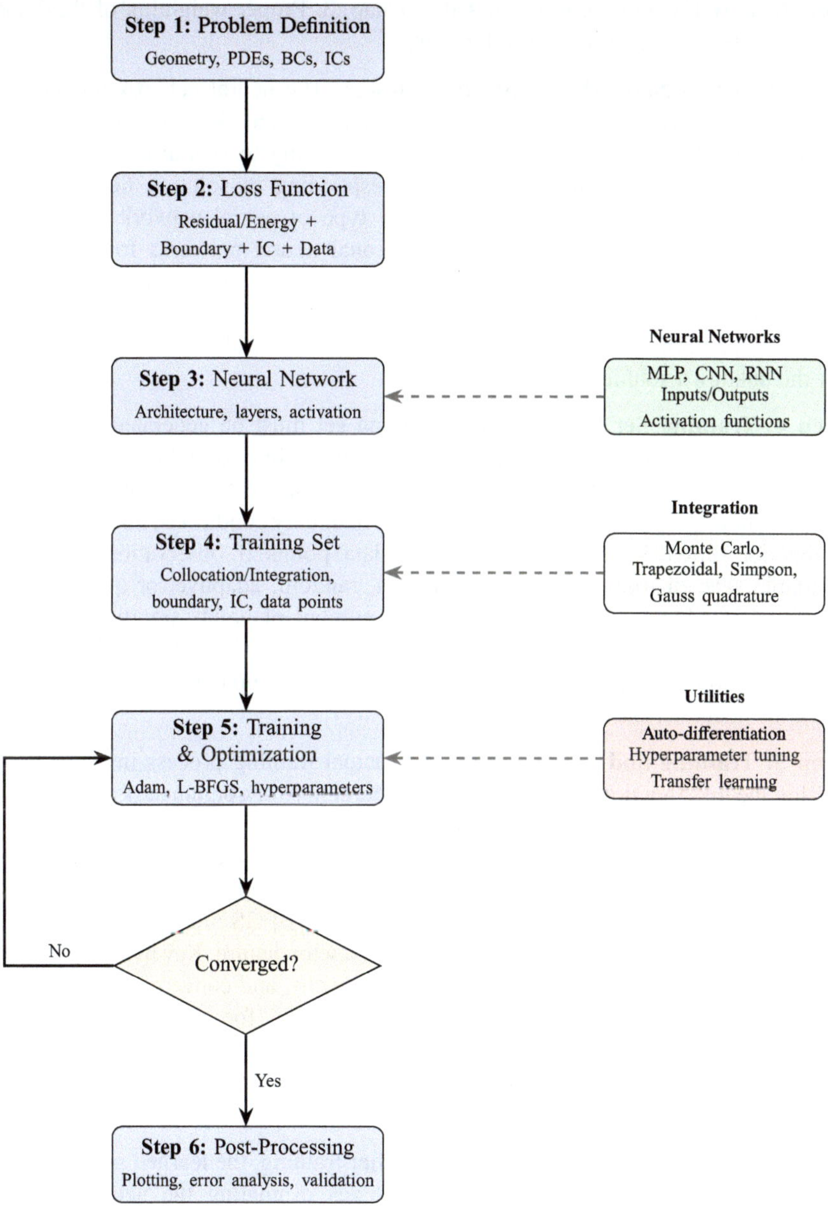

Fig. 2.20 Flowchart of the scientific machine learning implementation framework for engineering applications. The main workflow (Steps 1–6) is shown in the central column, with supporting modules indicated on the right

form (energy functional), as discussed in Chap. 3. Proper weighting of these loss components is crucial for balanced training.

Step 3: Neural Network Architecture Design The neural network architecture must be carefully designed to suit the problem at hand. Key decisions include: the number of input neurons (typically corresponding to spatial coordinates and time), the number of output neurons (corresponding to solution fields such as displacement, temperature, or pressure), the type of neural network (multi-layer perceptrons for general problems, convolutional neural networks for image-like data, recurrent neural networks for sequential data), the number of hidden layers and neurons per layer, and the choice of activation functions (e.g., tanh, sigmoid, ReLU, or specialized functions). The network serves as a universal function approximator for the unknown solution.

Step 4: Training Set Generation A training set must be generated to evaluate the loss function during optimization. This includes: interior collocation points or quadrature points distributed throughout the domain Ω for evaluating the PDE residual, boundary points on $\partial\Omega$ for enforcing boundary conditions, initial condition points for time-dependent problems, and data points if observational data are available. The distribution strategy (uniform, random, adaptive, or quasi-random such as Latin hypercube sampling) and the number of points significantly affect convergence and accuracy. Integration schemes such as Monte Carlo, Trapezoidal rule, Simpson's rule, or Gauss quadrature may be employed when computing integrals in energy-based formulations.

Step 5: Training and Optimization The actual training process involves minimizing the loss function with respect to the neural network parameters (weights and biases). This typically employs gradient-based optimizers such as Adam (adaptive moment estimation) or L-BFGS (limited-memory Broyden–Fletcher–Goldfarb–Shanno), as discussed in Sect. 2.3. Adam is commonly used for initial training due to its computational efficiency and robustness, while L-BFGS often provides superior convergence in the final stages due to its quasi-Newton nature. Key hyperparameters include learning rates, batch sizes, number of epochs, and convergence tolerances. Automatic differentiation (see Sect. 2.1) is essential for efficiently computing gradients. Hyperparameter optimization techniques (see Sect. 2.5) and transfer learning strategies (see Sect. 2.4) can be employed to enhance training efficiency and model performance.

Step 6: Post-Processing and Validation After training, the learned solution must be post-processed and validated. This involves evaluating the neural network at arbitrary points in the domain (which may differ from the training points) to visualize the solution fields, computing derived quantities such as stresses, fluxes, or gradients, estimating errors by comparing with analytical solutions or high-fidelity numerical simulations when available, and verifying that boundary and initial conditions are satisfied to acceptable tolerance. Visualization tools for plotting solution contours, deformation fields, or time evolution are essential for interpretation and communication of results.

The flowchart in Fig. 2.20 illustrates how these steps are logically connected in the implementation workflow. The modular structure allows flexibility in adapting the framework to different types of engineering problems while maintaining a consistent computational pipeline.

References

1. Bartholomew-Biggs, M., Brown, S., Christianson, B., Dixon, L.: Automatic differentiation of algorithms. J. Comput. Appl. Math. **124**(1–2), 171–190 (2000). 10.1016/S0377-0427(00)00422-2
2. Griewank, A., Walther, A.: Evaluating Derivatives: Principles and Techniques of Algorithmic Differentiation, 2nd edn. Society for Industrial and Applied Mathematics, Philadelphia (2008). ISBN: 9780898717761. 10.1137/1.9780898717761
3. Margossian, C.C.: A review of automatic differentiation and its efficient implementation. Wiley Interdiscip. Rev. Data Min. Knowl. Discov. **9**(4), e1305 (2019). 10.1002/widm.1305
4. Linnainmaa, S.: Taylor expansion of the accumulated rounding error. BIT Numer. Math. **16**(2), 146–160 (1976). 10.1007/BF01931367
5. Korelc, J., Wriggers, P.: Automation of Finite Element Methods. Springer, Berlin (2016). 10.1007/978-3-319-39005-5
6. Naumann, U.: The Art of Differentiating Computer Programs: An Introduction to Algorithmic Differentiation, vol. 24. SIAM, Philadelphia (2012). 10.1137/1.9781611972078
7. van Merriënboer, B., Moldovan, D., Wiltschko, A.B.: Tangent: Automatic differentiation using source-code transformation for dynamically typed array programming. Preprint. arXiv:1809.09569 (2018). 10.48550/arXiv.1809.09569
8. Haftka, R.T., Gürdal, Z.: Elements of Structural Optimization, vol. 11. Springer Science & Business Media, New York (2012). 10.1007/978-94-015-7862-2
9. Newton, I., Whiteside, D.: The Mathematical Papers of Isaac Newton: Volume 3 1670–1673. At the University Press, Cambridge (1969). 10.1063/1.3022301 https://books.google.de/books?id=T-7rzgEACAAJ
10. Broyden, C.G.: The convergence of a class of double-rank minimization algorithms 1. General considerations. IMA J. Appl. Math. **6**(1), 76–90 (1970). 10.1093/imamat/6.1.76
11. Broyden, C.G.: The convergence of a class of double-rank minimization algorithms: 2. The new algorithm. IMA J. Appl. Math. **6**(3), 222–231 (1970). 10.1093/imamat/6.3.222
12. Fletcher, R.: A new approach to variable metric algorithms. Comput. J. **13**(3), 317–322 (1970). 10.1093/comjnl/13.3.317
13. Goldfarb, D.: A family of variable-metric methods derived by variational means. Math. Comput. **24**(109), 23–26 (1970). 10.1090/S0025-5718-1970-0258249-6
14. Shanno, D.F.: Conditioning of quasi-Newton methods for function minimization. Math. Comput. **24**(111), 647–656 (1970). 10.1090/S0025-5718-1970-0274029-X
15. Kingma, D.P., Ba, J.: Adam: A method for stochastic optimization. Preprint. arXiv:1412.6980 (2014). 10.48550/arXiv.1412.6980
16. Ramachandran, P., Zoph, B., Le, Q.V.: Searching for activation functions. Preprint. arXiv:1710.05941 (2017). 10.48550/arXiv.1710.05941
17. Clevert, D.-A., Unterthiner, T., Hochreiter, S.: Fast and accurate deep network learning by exponential linear units (ELUS). Preprint. arXiv:1511.07289 (2015). 10.48550/arXiv.1511.07289
18. Misra, D.: Mish: A self regularized non-monotonic activation function. Preprint. arXiv:1908.08681 (2019). 10.48550/arXiv.1908.08681
19. Yonekura, K., Maruoka, K., Tyou, K., Suzuki, K.: Super-resolving 2D stress tensor field conserving equilibrium constraints using physics-informed U-Net. Finite Elem. Anal. Des. **213**, 103852 (2023). 10.1016/j.finel.2022.103852

20. Kamali, A., Laksari, K.: Physics-informed UNets for discovering hidden elasticity in heterogeneous materials. J. Mech. Behav. Biomed. Mater. **150**, 106228 (2024). 10.1016/j.jmbbm.2023 106228.
21. He, K., Zhang, X., Ren, S., Sun, J.: Deep residual learning for image recognition. In: Proceedings of the IEEE Conference on Computer Vision and Pattern Recognition, pp. 770–778 (2016). 10.1109/CVPR.2016.90
22. E, W., Bing, Y.: The deep Ritz method: a deep learning-based numerical algorithm for solving variational problems. Commun. Math. Stat. **6**(1), 1–12 (2018). 10.1007/s40304-018-0127-z
23. Wang, S., Li, B., Chen, Y., Perdikaris, P.: PirateNets: physics-informed deep learning with residual adaptive networks. J. Mach. Learn. Res. **25**(402), 1–51 (2024). http://jmlr.org/papers/v25/24-0313.html
24. Zhuang, F., Qi, Z., Duan, K., Xi, D., Zhu, Y., Zhu, H., et al.: A comprehensive survey on transfer learning. Proc. IEEE **109**(1), 43–76 (2020). 10.1109/JPROC.2020.3004555
25. Xu, C., Cao, B.T., Yuan, Y., Meschke, G.: Transfer learning based physics-informed neural networks for solving inverse problems in engineering structures under different loading scenarios. Comput. Methods Appl. Mech. Eng. **405**, 115852 (2023). 10.1016/j.cma.2022.115852
26. Guo, H., Zhuang, X., Chen, P., Alajlan, N., Rabczuk, T.: Analysis of three-dimensional potential problems in non-homogeneous media with physics-informed deep collocation method using material transfer learning and sensitivity analysis. Eng. Comput. **38**(6), 5423–5444 (2022). 10.1007/s00366-022-01633-6
27. Chakraborty, A., Anitescu, C., Zhuang, X., Rabczuk, T.: Domain adaptation based transfer learning approach for solving PDEs on complex geometries. Eng. Comput. **38**(5), 4569–4588 (2022). 10.1007/s00366-022-01661-2
28. Wang, Y., Bai, J., Lin, Z., Wang, Q., Anitescu, C., Sun, J., et al.: Artificial intelligence for partial differential equations in computational mechanics: a review. Preprint. arXiv:2410.19843 (2024). 10.48550/arXiv.2410.19843
29. He, J., Zhou, C., Ma, X., Berg-Kirkpatrick, T., Neubig, G.: Towards a unified view of parameter-efficient transfer learning. Preprint. arXiv:2110.04366 (2021). 10.48550/arXiv.2110.04366
30. Radford, A., Wu, J., Child, R., Luan, D., Amodei, D., Sutskever, I., et al.: Language models are unsupervised multitask learners. OpenAI Blog **1**(8), 9 (2019). https://cdn.openai.com/better-language-models/language_models_are_unsupervised_multitask_learners.pdf
31. Goswami, S., Anitescu, C., Chakraborty, S., Rabczuk, T.: Transfer learning enhanced physics informed neural network for phase-field modeling of fracture. Theoret. Appl. Fract. Mech. **106**, 102447 (2020). 10.1016/j.tafmec.2019.102447
32. Chakraborty, S.: Transfer learning based multi-fidelity physics informed deep neural network. J. Comput. Phys. **426**, 109942 (2021). 10.1016/j.jcp.2020.109942
33. Hu, E.J., Shen, Y., Wallis, P., Allen-Zhu, Z., Li, Y., Wang, S., et al.: Lora: Low-rank adaptation of large language models. Preprint. arXiv:2106.09685 (2021). 10.48550/arXiv.2106.09685
34. Wang, Y., Bai, J., Eshaghi, M.S., Anitescu, C., Zhuang, X., Rabczuk, T., et al.: Transfer learning in physics-informed neurals networks: full fine-tuning, lightweight fine-tuning, and low-rank adaptation. Int. J. Mech. Syst. Dyn. (2025). 10.1002/msd2.70030
35. Shrestha, A., Mahmood, A.: Review of deep learning algorithms and architectures. IEEE Access **7**, 53040–53065 (2019). 10.1109/ACCESS.2019.2912200
36. Thornton, C., Hutter, F., Hoos, H.H., Leyton-Brown, K.: Auto-WEKA: Combined selection and hyperparameter optimization of classification algorithms. In: Proceedings of the 19th ACM SIGKDD International Conference on Knowledge Discovery and Data Mining, pp. 847–855 (2013). 10.1145/2487575. 2487629
37. Yao, Q., Wang, M., Chen, Y., Dai, W., Li, Y.-F., Tu, W.-W., et al.: Taking human out of learning applications: a survey on automated machine learning. Preprint. arXiv:1810.13306 (2018). 10.48550/arXiv.1810.13306
38. Elsken, T., Metzen, J.H., Hutter, F.: Neural architecture search: a survey. J. Mach. Learn. Res. **20**(1), 1–21 (2019). http://jmlr.org/papers/v20/18-598.html

39. Ren, P., Xiao, Y., Chang, X., Huang, P.-Y., Li, Z., Chen, X., et al.: A comprehensive survey of neural architecture search: challenges and solutions. ACM Comput. Surv. (CSUR) **54**(4), 1–34 (2021). 10.1145/3447582
40. Real, E., Aggarwal, A., Huang, Y., Le, Q.V.: Regularized evolution for image classifier architecture search. In: Proceedings of the AAAI Conference on Artificial Intelligence, vol. 33, pp. 4780–4789 (2019). 10.1609/aaai.v33i01.33014780
41. Liu, H., Simonyan, K., Yang, Y.: DARTS: Differentiable architecture search. Preprint. arXiv:1806.09055 (2018). 10.48550/arXiv.1806.09055
42. Liu, C., Zoph, B., Neumann, M., Shlens, J., Hua, W., Li, L.-J., et al.: Progressive neural architecture search. In: Proceedings of the European Conference on Computer Vision (ECCV), pp. 19–34 (2018). 10.1007/978-3-030-01246-52
43. Pham, H., Guan, M.Y., Zoph, B., Le, Q.V., Dean, J.: Efficient neural architecture search via parameter sharing. Preprint. arXiv:1802.03268 (2018). 10.48550/arXiv.1802.03268
44. Guo, H., Zhuang, X., Chen, P., Alajlan, N., Rabczuk, T.: Stochastic deep collocation method based on neural architecture search and transfer learning for heterogeneous porous media. Eng. Comput. **38**(6), 5173–5198 (2022). 10.1007/s00366-021-01586-2
45. Guo, H., Zhuang, X., Alajlan, N., Rabczuk, T.: Physics-informed deep learning for melting heat transfer analysis with model-based transfer learning. Comput. Math. Appl. **143**, 303–317 (2023). 10.1016/j.camwa.2023.05.014
46. Wang, Y., Zhong, L.: NAS-PINN: neural architecture search-guided physics-informed neural network for solving PDEs. J. Comput. Phys. **496**, 112603 (2024). 10.1016/j.jcp.2023.112603
47. Ruchte, M., Zela, A., Siems, J.N., Grabocka, J., Hutter, F.: NASLib: a modular and flexible neural architecture search library (2020). https://openreview.net/references/pdf?id=Ja_N_68HG
48. Pestourie, R., Mroueh, Y., Rackauckas, C., Das, P., Johnson, S.G.: Physics-enhanced deep surrogates for partial differential equations. Nat. Mach. Intell., 1–8 (2023). 10.1038/s42256-023-00761-y
49. Hennigh, O., Narasimhan, S., Nabian, M.A., Subramaniam, A., Tangsali, K., Fang, Z., et al.: "NVIDIA SimNet": an AI-accelerated multi-physics simulation framework. In: International Conference on Computational Science, pp. 447–461. Springer (2021). 10.1007/978-3-030-77977-136
50. Lu, L., Meng, X., Mao, Z., Karniadakis, G.E.: DeepXDE: a deep learning library for solving differential equations. SIAM Rev. **63**(1), 208–228 (2021). 10.1137/19M1274067
51. Peng, W., Zhang, J., Zhou, W., Zhao, X., Yao, W., Chen, X.: IDRLnet: a physics-informed neural network library. Preprint. arXiv:2107.04320 (2021). 10.48550/arXiv.2107.04320
52. Chen, F., Sondak, D., Protopapas, P., Mattheakis, M., Liu, S., Agarwal, D., et al.: Neurodiffeq: a python package for solving differential equations with neural networks. J. Open Source Softw. **5**(46), 1931 (2020). 10.21105/joss.01931
53. Lu, L., Jin, P., Pang, G., Zhang, Z., Karniadakis, G.E.: Learning nonlinear operators via DeepONet based on the universal approximation theorem of operators. Nat. Mach. Intell. **3**(3), 218–229 (2021). 10.1038/s42256-021-00302-5
54. Li, Z., Kovachki, N., Azizzadenesheli, K., Liu, B., Bhattacharya, K., Stuart, A., et al.: Fourier neural operator for parametric partial differential equations. Preprint. arXiv:2010.08895 (2020). 10.48550/arXiv.2010.08895
55. Goswami, S., Yin, M., Yu, Y., Karniadakis, G.E.: A physics-informed variational DeepONet for predicting crack path in quasi-brittle materials. Comput. Methods Appl. Mech. Eng. **391**, 114587 (2022). 10.1016/j.cma.2022.114587
56. Tan, L., Chen, L.: Enhanced DeepONet for modeling partial differential operators considering multiple input functions. Preprint. arXiv:2202.08942 (2022). 10.48550/arXiv.2202.08942
57. He, J., Kushwaha, S., Park, J., Koric, S., Abueidda, D., Jasiuk, I.: Sequential Deep Operator Networks (S-DeepONet) for predicting full-field solutions undertime-dependent loads. Eng. Appl. Artif. Intell. **127**, 107258 (2024). 10.1016/j.engappai.2023.107258
58. Lu, L., Meng, X., Cai, S., Mao, Z., Goswami, S., Zhang, Z., et al.: A comprehensive and fair comparison of two neural operators (with practical extensions) based on FAIR data. Comput. Methods Appl. Mech. Eng. **393**, 114778 (2022). 10.1016/j.cma.2022.114778

59. Wen, G., Li, Z., Azizzadenesheli, K., Anandkumar, A., Benson, S.M.: U-FNO - an enhanced Fourier neural operator-based deep-learning model for multiphase flow. Adv. Water Resour. **163**, 104180 (2022). 10.1016/j.advwatres.2022.104180
60. You, H., Zhang, Q., Ross, C.J., Lee, C.-H., Yu, Y.: Learning deep implicit Fourier neural operators (IFNOs) with applications to heterogeneous material modeling. Comput. Methods Appl. Mech. Eng. **398**, 115296 (2022). 10.1016/j.cma.2022.115296
61. Pathak, J., Subramanian, S., Harrington, P., Raja, S., Chattopadhyay, A., Mardani, M., et al.: FourCastNet: a global data-driven high-resolution weather model using adaptive Fourier Neural Operators. Preprint. arXiv:2202.11214 (2022). 10.48550/arXiv.2202.11214
62. McIntosh, T.R., Susnjak, T., Liu, T., Watters, P., Xu, D., Liu, D., et al.: From Google Gemini to OpenAI Q*(Q-star): a survey on reshaping the generative artificial intelligence (AI) research landscape. Technologies **13**(2), 51 (2025)
63. Kumar, V., Gleyzer, L., Kahana, A., Shukla, K., Karniadakis, G.E.: My-CrunchGPT: a chatGPT assisted framework for scientific machine learning. Preprint. arXiv:2306.15551 (2023). 10.48550/arXiv.2306.15551
64. Cohen, S.N., Jiang, D., Sirignano, J.: Neural Q-learning for solving PDEs. J. Mach. Learn. Res. **24**(236), 1–49 (2023). http://jmlr.org/papers/v24/22-1075.html

Chapter 3
Partial Differential Equations in Engineering

3.1 Introduction

Differential equations have first been introduced in the seventeenth century by Isaac Newton and Gottfried Leibniz though the foundation has been laid by the French mathematician Guillaume de l'Hopital years before. The advent of differential equations as physical models has been laid in the eighteenth century by Euler, d'Alembert, Lagrange and Laplace who focused on the mechanics of continua. Partial Differential Equations (PDEs) are fundamental to model different phenomena in science, physics and engineering mathematically. Though the underlying physics/phenomena might be different depending on the application or the research area, the underlying PDE might share some similarities inherent to its 'nature', which we will discuss later. For now, let us focus on a specific example and consider the PDE for a potential function ϕ

$$(k(\boldsymbol{x})\phi_{,i})_{,i} = k(\boldsymbol{x})\phi_{,ii} + k_{,i}(\boldsymbol{x})\phi_{,i} = 0 \tag{3.1}$$

where $k(\boldsymbol{x})$ is a function which depends on the spatial coordinates $\boldsymbol{x}$ (or alternatively on material coordinates $\mathbf{X}$) and which has different physical meaning—such as heat conductivity, permeability, permittivity, resistivity, magnetic permeability—depending on the application. Equation (3.1) is a 'special case' of the convection-diffusion equation (stationary diffusion equation) and the governing equation for a wide range of problems in physics and engineering such as heat transfer, incompressible flow, gravity fields, shaft torsion, electrostatics and magnetostatics, see also Table 3.1 [1]. The convection-diffusion equation—in general—mathematically describes transport phenomena, which have countless applications in engineering, physics and mechanics.

Let us consider Newton's equation of motion as another example. It is the governing equation in molecular dynamics (MD), determining the trajectories

T. Rabczuk et al., *Scientific Machine Learning with Engineering Applications*,
Studies in Systems, Decision and Control 667,
https://doi.org/10.1007/978-3-032-20307-6_3

Table 3.1 Problems belong to the category of problems of potential

Problems	Scalar function ϕ	$k(\boldsymbol{x})$	Boundary condition	
			Dirichlet	Neumann
Heat transfer	Temperature T	Thermal conductivity (k)	$T = \bar{T}$	Heat flow $q = -k\frac{\partial T}{\partial n}$
Ground water flow	Hydraulic head H	Permeability (k)	$H = \bar{H}$	Velocity flow $q = -k\frac{\partial H}{\partial n}$
Electrostatic	Field potential V	Permittivity (ε)	$V = \bar{V}$	Electric flow $q = -k\frac{\partial V}{\partial n}$
Electric conduction	Electropotential E	Conductivity (k)	$E = \bar{E}$	Electric current $q = -k\frac{\partial E}{\partial n}$
Magnetostatic	Magnetic potential M	Magnetic permeability (μ)	$M = \bar{M}$	Magnetic flux density $q = -k\frac{\partial M}{\partial n}$

of molecules or atoms. The interactions between these molecules and atoms is determined by so-called interatomic potentials, which are found and calibrated through DFT simulations or experiments at very small length scales, commonly in the order of nanometers. The forces exchanging impulses between atoms are simply the derivatives of the interatomic potential with respect to (w.r.t.) the position vector of the atoms or molecules. MD simulations have been extensively used in physics and materials science to better understand certain physical phenomena or to predict macroscopic properties of materials such as strength, thermal conductivity or other electrical, optical, thermal or mechanical quantities. On the other hand, Newton's equation of motion is also the basis for many problems in solid mechanics with key applications in engineering. However, in contrast to MD, the material is considered as a continuum and the 'internal force vector' is related to a specific stress tensor, for instance the divergence of the Cauchy stress tensor. In this case, the governing equation is known as the Cauchy momentum equation. The Navier Stokes equations with applications in fluid dynamics also stems from Cauchy momentum equation, which in turn shows some similarities to the Black-Scholes-Merton PDE used in financial mathematics in order to predict the behaviour of financial markets. Though the solution of PDEs are relevant in many disciplines, the key focus of this book will be on applications in engineering, mainly in solid mechanics.

Second-order PDEs—such as the Cauchy momentum equation or the convection-diffusion equation—are among the most common ones in engineering and mechanics. As the name suggests they require second-order partial derivatives of the primary variables and can be classified into elliptic, parabolic and hyperbolic. The classification of PDEs provides a guide to appropriate initial and boundary conditions and to the smoothness of the solutions. Proving the smoothness of a solution for a specific application is not always straightforward. For instance, while a proof for smooth solutions to the two-dimensional Navier-Stokes equation was given in the 1960s [2, 3], there is no mathematical proof yet for the existence of smooth solutions to the three dimensional Navier Stokes problem. In fact, the problem of Navier Stokes smoothness and existence is one of the millennium prize problems which is rewarded with a 1,000,000 USD prize. Let us now consider a scalar-valued linear second-order partial differential equation (PDE) in 3D given by

$$A_{ij}u(x_i)_{,ij} = f(x_i, u(x_i)_{,i}, u(x_i)) \tag{3.2}$$

where the comma indicates partial derivatives, i.e. $()_{,i} = \frac{\partial ()}{\partial x_i}$, x_i being the spatial (or alternatively material X_i) coordinates and A_{ij} are the coefficients of a 3×3 matrix. The right hand side of Eq. (3.2) contains only terms of lower order derivatives. Note that we used Einstein's sum notation and sum over repeated indices. It can then be shown that the PDE is elliptic if the eigenvalues of the coefficient matrix are all negative or all positive. If the coefficient matrix has both negative *and* positive eigenvalues, the PDE is hyperbolic and if at least one eigenvalue is zero, the PDE is parabolic. Note that the coefficients A_{ij} are not constant for many engineering applications (but functions of x_i) and the type of the PDE can change for a single problem at hand, for instance when the sign of the coefficients change. A classical example are material instabilities, that finally lead to fracture. In static applications, the governing equation is the equilibrium equation, which is a special case of the equation of motion, namely when the inertia term is zero. Note that the equilibrium equation for a linear elastic solid is an elliptic PDE while the equation of motion (with a non-zero inertia term) is hyperbolic. However, for non-linear material behaviour for instance, the coefficients are not constant but depend on the deformation or strain state and the loading history. Many solids exhibit a so-called strain softening, so that the coefficients become negative which changes the type of the PDE. Note also that the classification of PDE in the literature is focused on second-order scalar-valued PDEs. However, for higher-order PDEs or vector-valued PDEs, the classification of PDEs is more complicated and so is the associated proof of well-posedness, which we address later. Furthermore, there are several PDEs, which do not fall in any of above categories (elliptic, parabolic or hyperbolic).

There are also several relevant higher-order PDEs in engineering. In structural analysis for instance, Euler-Bernoulli beam theory and its 'extension' to plates (Kirchhoff plate) and shells (Kirchhoff-Love shell) are fourth-order PDEs, which stem from a constraint condition, i.e. requesting that the director field needs to remain perpendicular to the mid-surface of the plate/shell/beam, which finally relates an angle to the deflection (actually its derivatives) such that the final PDE can be expressed solely in terms of the deflection or more precisely, its fourth-order derivatives. Also higher-order shear deformation theory yields higher order PDEs. Other interesting examples of higher-order PDEs include flexoelectricity or higher-order phase field models which govern many phenomena in physics and engineering, though the strong form for these applications is commonly derived from an energy functional. We will come back to that later on. Many relevant problems in engineering require the solution of coupled PDEs, which are based on several primary variables. It is for instance well known that the temperature is governing the propagation of shear bands leading to a coupled thermo-mechanical problem. Energy harvesters transferring mechanical to electrical energy is another engineering problem, which requires a solution of coupled PDEs.

In order to obtain a—unique—solution for a specific problem of interest, the PDE alone is not enough. It needs to be complemented by boundary (and initial conditions for transient problems) which yields the so-called boundary value problem (or initial boundary value problem for transient problems). There are three type of boundary conditions, i.e. Dirichlet boundary conditions, Neumann boundary condition and Robin boundary conditions. At the Dirichlet boundary, values of the primary field itself are imposed while Neumann boundary conditions impose values of its derivatives. Robin boundary conditions are a weighted combination of the first two and less common in engineering applications. Many PDE problems are well-posed, meaning they have a solution, its solution is unique and stable (meaning that a small change in the data results in a small change in the solution). However, there are also interesting problems, which are ill-posed; material instabilities and fracture mentioned previously is one example. Other examples are buckling and inverse or optimization problems which are crucial in monitoring and designing new materials and structures.

Solving the BVP (or IBVP) is a crucial step towards a precise knowledge of the behavior of natural and engineered systems. In order to solve a specific BVP (or IBVP) that represents real systems to an acceptable degree, analytical methods are usually not enough and one has to resort to discretization methods. Finite Difference Methods (FDM), Finite Volume Methods (FVM), Boundary Element Methods (BEM), Finite Element Methods (FEM), Meshfree Methods (MMs) and Isogeometric Analysis (IGA) are among the most common discretization methods used in engineering and mechanics, just to name a few. While the FDM solves the BVP in strong form, meaning at discrete points, the FEM, which is probably the most popular and most frequently used discretization method in engineering, is based on the weak form. The weak form can be easily derived from the strong form and has lower requirements on the differentiability of the primary variable. Also the fulfillment of Neumann boundary conditions is simpler. And though generalized FDMs allow for unstructured discretization, one key advantage of FEM is its ability to easily model complex three-dimensional geometries. Its popularity is 'proven' through numerous commercial (ABAQUS, ANSYS just to name two) and open-source FEM packages. Most IGA formulations are based on the weak form though there are a few contributions where a strong form has been exploited. IGA was developed in 2005 by Hughes et al. [4] with the objective to unify Computer Aided Design (CAD) and Computer Aided Engineering (CAE) to finally drastically accelerate the design-to-analysis process. This is achieved by using the same basis functions for design and analysis. While FEM is commonly based on Lagrange polynomials, IGA exploits spline functions such as B-splines, Non-Uniform Rational B-Splines (NURBS), T-Splines, PHT- or THB-splines, to name a few. They have commonly been used in CAD for surface parametrizations and are constructed through tensor-products in higher order dimensions. Volume parametrizations and the generation of so-called analysis suitable geometries still remains a major challenge in IGA. For instance, complex geometries often contain gaps and overlaps, which requires repairing. However, CAD basis functions such as splines have shown superiority to Lagrange polynomials in many applications including dynamics [5] and due to their higher-order continuity also for higher-order PDEs [6]. Meshfree methods (MMs) are based either on the strong form or the weak form. They have disadvantages in surface parametrizations and advantages in volume parametrizations. For that reason, MMs have been coupled with FEM [7–9] and also IGA [10, 11]. The BEM reduces the dimension of the problem by one order as it requires only the discretization of the boundary. However, BEM is not computationally more efficient than FEM as the system matrix is not sparse. Furthermore, BEM requires information of the solution in the interior significantly limiting its application. For that reason, BEM is not a popular alternative in the engineering and mechanics community.

Consider now a generalized PDE expressed as:

$$\mathcal{N}[\boldsymbol{u}] = 0 \quad \text{on} \quad \Omega, \tag{3.3}$$

where $\mathcal{N}$ denotes a (possibly non-linear) differential operator acting upon a (possibly vector-valued) function, $\boldsymbol{u}$. In addition, $\boldsymbol{u}$ must satisfy both the Dirichlet and the Neumann-boundary conditions. As mentioned before, the solution of this PDE subjected to appropriate boundary conditions is referred to as a boundary value problem (BVP). Many problems in science and engineering can be written as specific cases of this generalized format. Sometimes, this version of the corresponding BVP is called the strong form. Let us apply this general setting to the boundary value problem of linear elastostatic for a homogeneous isotropic solid. The first step is the definition of a primary variable. We opt to choose the displacement field $\boldsymbol{u}(\boldsymbol{x})$. Then, the linear strain tensor $\boldsymbol{\epsilon}(\boldsymbol{u}(\boldsymbol{x}))$ is defined as follows:

$$\boldsymbol{\epsilon}(\boldsymbol{u}(\boldsymbol{x})) = \frac{1}{2}\Big(\nabla \boldsymbol{u} + (\nabla \boldsymbol{u})^T\Big). \tag{3.4}$$

$\nabla = \frac{\partial}{\partial x}$ indicating the nabla operator. The Cauchy stress tensor, $\boldsymbol{\sigma}(\boldsymbol{\epsilon})$ can be computed through Hooke's law by

$$\boldsymbol{\sigma}(\boldsymbol{\epsilon}) = \mathbb{C} : \boldsymbol{\epsilon}, \tag{3.5}$$

where $\mathbb{C}$ denotes the fourth-order elasticity tensor. Taking advantage of conservation of angular momentum, it can be shown that the Cauchy-stress tensor is symmetric. The equilibrium equation can be derived at an infinitesimal volume element through force equilibrium, which is finally expressed in terms of the components of the Cauchy-stress tensor given by

$$\nabla \cdot \boldsymbol{\sigma} + \mathbf{b} = \mathbf{0}. \tag{3.6}$$

with **b** indicating body forces. Hence, $\mathcal{N}[\boldsymbol{u}]$ from Eq. (3.3) is

$$\mathcal{N}[\boldsymbol{u}] = \nabla \cdot \boldsymbol{\sigma}\big(\boldsymbol{\epsilon}(\boldsymbol{u})\big). \tag{3.7}$$

By substituting Eq. (3.4) into Eq. (3.5) and its outcome in (3.6), the equilibrium equation can be expressed directly in terms of the primary variables, i.e. the displacement field **u**:

$$\lambda u_{k,ki} + \nu\left(u_{i,jj} + u_{j,ij}\right) + b_i = 0 \tag{3.8}$$

which is also known as the Navier-Lamé equation where λ and ν denote the Lamé constants. These equations need to be complemented by boundary conditions given by

$$\begin{aligned} \mathbf{u} &= \bar{\mathbf{u}} \text{ on } \Gamma_u \\ \boldsymbol{\sigma} \cdot \mathbf{n} = \mathbf{t} &= \bar{\mathbf{t}} \text{ on } \Gamma_t, \end{aligned} \tag{3.9}$$

Γ_u and Γ_t referring to the Dirichlet and von Neumann boundaries with $\Gamma_u \bigcup \Gamma_t = \Gamma$ and $\Gamma_u \bigcap \Gamma_t = \emptyset$ and where the superimposed bar indicates the imposed boundary conditions; **t** is the stress vector, also called traction, and **n** the outer normal to the boundary Γ. The first term on the LHS in the second line of Eq. (3.9) is the so-called Cauchy theorem relating the stress vector to the stress tensor. In summary, we have 15 unknowns and 15 equations defined in the domain Ω (in 3D), i.e. six components of the Cauchy stress tensor $\boldsymbol{\sigma}$, 6 components of the linear strain tensor $\boldsymbol{\epsilon}$ and the 3 components of the displacement vector **u**. The relationship between these components can be illustrated in a Tonti diagram as depicted in Fig. 3.1, which is a powerful and illustrative 'tool' that can be applied to any BVP to visualize the relationships between all variables. As mentioned before, the analytical solution of the BVP for generic geometries is difficult. For sufficiently smooth $\boldsymbol{u}$, it can however easily be solved by a collocation-type method. As Eq. (3.8) clearly shows, this requires that the primary variable is two times continuous differentiable.

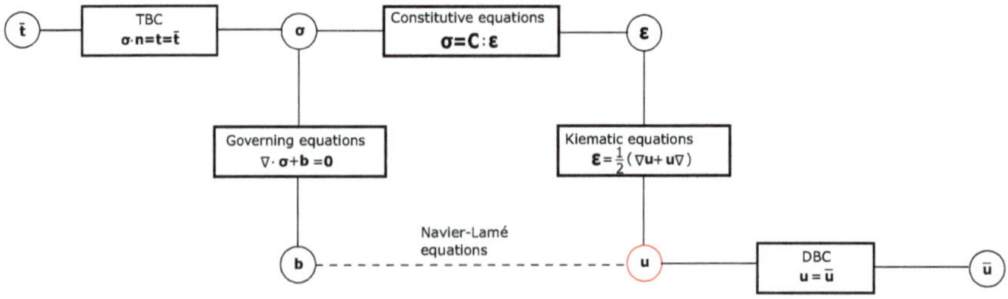

Fig. 3.1 Tonti-diagram of the boundary value problem for linear elastostatics and a pure displacement based formulation in the strong form

The weak form, which is the basis for many discretization methods including the FEM, can be derived from the strong form (and vice versa) using the method of weighted residuals. The first step is to define the equations to be fulfilled in the weak form, see the Tonti diagram in Fig. 3.1. Commonly the equilibrium equation and the Neumann boundary conditions are enforced in weak form while Dirichlet boundary conditions are imposed in strong form, at least in most finite element formulations. Then, the corresponding equations are multiplied with a test function $\delta\mathbf{u}$ and integrated over a control volume yielding

$$\int_{\Omega} \delta\mathbf{u} \cdot (\nabla \cdot \boldsymbol{\sigma} + \mathbf{b}) \; d\Omega + \int_{\Gamma} \delta\mathbf{u} \cdot \left(\boldsymbol{\sigma} \cdot \mathbf{n} - \bar{\mathbf{t}}\right) \; d\Gamma = 0 \tag{3.10}$$

Next, we consider the first part of the domain integral $\int_{\Omega} \delta\mathbf{u} \cdot \nabla \cdot \boldsymbol{\sigma} \; d\Omega$ and employ integration by parts yielding

$$\int_{\Omega} \delta\mathbf{u} \cdot \nabla\boldsymbol{\sigma} \; d\Omega = \int_{\Omega} \nabla \cdot (\delta\mathbf{u} \cdot \boldsymbol{\sigma}) \; d\Omega - \int_{\Omega} \nabla\delta\mathbf{u} : \boldsymbol{\sigma} \; d\Omega \tag{3.11}$$

The first term on the RHS of Eq. (3.11) can be transformed into a boundary integral using Gauss divergence theorem:

$$\int_{\Omega} \delta\mathbf{u} \cdot \nabla \cdot \boldsymbol{\sigma} \; d\Omega - \int_{\Gamma} (\delta\mathbf{u} \cdot \boldsymbol{\sigma}) \cdot \mathbf{n} \, d\Gamma - \int_{\Omega} \delta\epsilon : \boldsymbol{\sigma} \; d\Omega \tag{3.12}$$

Note that the Gauss divergence theorem requires the integrand in the brackets of the boundary integral to be C^0 (zero times continuous differentiable), which is for instance not fulfilled in finite elements which have jumps in the stresses at the element boundaries; the quantities in the jumps are often used as indicators for h-adaptive refinements. Let us consider now the integrand of the last term on the LHS $\nabla\delta\mathbf{u} : \boldsymbol{\sigma}$. The gradient of the test function $\nabla\delta\mathbf{u}$ is generally an unsymmetric second-order tensor, that can be additively decomposed into a symmetric and antisymmetric one: $\nabla\delta\mathbf{u} = (\nabla\delta\mathbf{u})^{symm} + (\nabla\delta\mathbf{u})^{asymm}$ or $\nabla\delta\mathbf{u} = \delta\epsilon + \delta\boldsymbol{\omega}$; the symmetric part of the displacement gradient of the test function can also be interpreted as the virtual strain tensor (in the context of the principle of virtual work). As the second-order tensor product between a symmetric tensor (such as the Cauchy-stress tensor) and an antisymmetric tensor (as the antisymmetric part of the gradient of the test function) vanishes, Eq. (3.12) can be rewritten as

$$\int_{\Omega} \delta\mathbf{u} \cdot \nabla\boldsymbol{\sigma} \; d\Omega = \int_{\Gamma} (\delta\mathbf{u} \cdot \boldsymbol{\sigma}) \cdot \mathbf{n} \, d\Gamma - \int_{\Omega} \delta\epsilon : \boldsymbol{\sigma} \; d\Omega \tag{3.13}$$

Substituting Eq. (3.13) into Eq. (3.11) leads to the final weak form

$$\int_{\Gamma_t} \delta\mathbf{u} \cdot \bar{\mathbf{t}} \, d\Gamma + \int_{\Omega} \delta\mathbf{u} \cdot \mathbf{b} \, d\Omega - \int_{\Omega} \delta\epsilon : \boldsymbol{\sigma} \; d\Omega = 0 \tag{3.14}$$

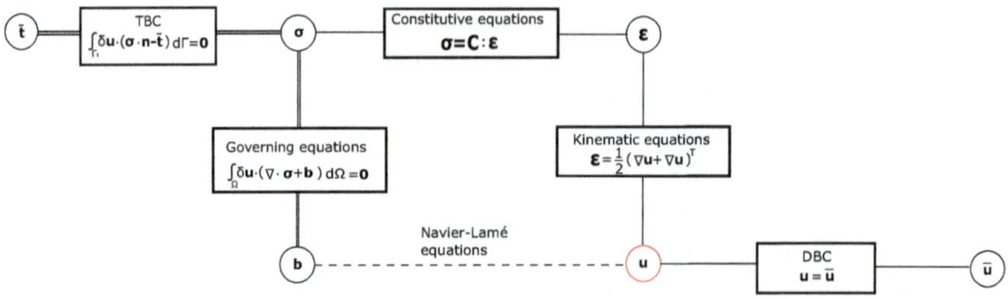

Fig. 3.2 Tonti-diagram of the boundary value problem for linear elastostatics and a pure displacement based formulation in the weak form

The associated Tonti diagram for the weak form is illustrated in Fig. 3.2. The solid lines indicate that the associated equations are imposed in strong form while the dashed lines indicate the weak form. Note that the weak form can be obtained also through an energy approach assuming a suitable energy can be defined, which will be discussed next.

For the energy approach to a BVP, consider the following problem

$$\min_{\boldsymbol{u}} \mathcal{E}[\boldsymbol{u}], \tag{3.15}$$

where $\boldsymbol{u}$ is constrained by Dirichlet boundary conditions. Here, we will assume that the total energy of the system, $\mathcal{E}$, is such that there exists a unique solution of the problem defined in (3.15) and is the same as the solution of (3.3). In that case, (3.3) is called the Euler-Lagrange equation of the variational problem defined in (3.15). The first step to go from the variational formulation to the corresponding strong form is to find a stationary state of $\mathcal{E}$ by equating its first variation to zero:

$$\delta\mathcal{E}[\boldsymbol{u}, \delta\boldsymbol{u}] = \frac{\mathrm{d}}{\mathrm{d}\tau}\mathcal{E}[\boldsymbol{u} + \tau\delta\boldsymbol{u}]\bigg|_{\tau=0} = 0, \tag{3.16}$$

which has to hold for any admissible $\delta\boldsymbol{u}$ (i.e., for any smooth enough function having homogeneous boundary conditions on the Dirichlet part of the boundary). The resulting expression is the so-called weak form of the problem defined by (3.16) plus appropriate boundary conditions. In mechanical problems, this corresponds to the principle of virtual work. As already pointed out, the weak form is the point of departure of important discretization methods such as the FEM. Once the weak form is established, the strong form can be derived by going through the steps described before but in reverse order. Note that the machine learning approach to the solution of a PDE deals directly with the minimization of the energy $\mathcal{E}$, circumventing the need to derive the weak form from the energy of the system explicitly. Many interesting engineering problems are based on an energy. We will come back to this in Sect. 3.4. However, let us illustrate this approach now again for a linear elastic isotropic solid. One can define the stored elastic strain energy, $\Psi(\boldsymbol{\epsilon})$ as:

$$\Psi(\boldsymbol{\epsilon}) = \frac{1}{2}\boldsymbol{\epsilon} : \mathbb{C} : \boldsymbol{\epsilon}. \tag{3.17}$$

Notice that the Cauchy stress tensor, $\boldsymbol{\sigma}(\boldsymbol{\epsilon})$, can be computed as follows:

$$\boldsymbol{\sigma}(\boldsymbol{\epsilon}) = \frac{\partial \Psi}{\partial \boldsymbol{\epsilon}}. \tag{3.18}$$

Then, for the case of zero body forces and homogeneous Neumann boundary conditions, we can define the energy of the solid as

$$\mathcal{E}[\boldsymbol{u}] = \int_{\Omega} \Psi\big(\boldsymbol{\epsilon}(\boldsymbol{u})\big) d\Omega. \tag{3.19}$$

Above formulation is based on a single field potential depending on the displacement field indicated by the Tonti diagram in Fig. 3.2. The strain and the stress field are then derived from this primary field. However, one can also choose several primary variables which yield in case of finite elements to mixed or hybrid formulations. The two-field Hellinger Reissner potential and the three-field Hu-Washizu potential are classical examples.

The energy approach to a BVP has several advantages over collocation methods which are based on the strong form: First of all, it reduces the requirement on the differentiability of the primary field. Secondly, it automatically fulfills zero Neumann boundary conditions as above derivation (deriving the weak form from the strong form) illustrates. And finally, energy approaches are the basis for thermodynamic consistent formulations. It can be shown that the existence of a suitable energy guarantees the fulfillment of the conservation laws in continuum mechanics including the first and second law of thermodynamics, which play a major role in constitutive modeling as formulations based on an energy potential can avoid unphysical behaviour. Hyperelasticity or the thermodynamic consistent framework of the phase field model of fracture [12, 13] are classical examples that demonstrate the power of energy approaches. They are also related to the solution of non-linear PDEs, especially the concepts of entropy and convexity. Ball and Chen [14] for instance mention the severe restrictions of typical classifications of PDE and suggests using the notions of entropy (which is equivalent to using energies and dissipation) and convexity instead, which play also a major role in continuum mechanics and thermodynamics.

In the next section, we will focus on some—in engineering and mechanics popular—second-order PDEs, which we will later on solve with a collocation-based machine learning approach (in strong form), which are commonly referred to as Physics-Informed Neural Networks (PINNs). We subsequently present some energy approaches to specific BVP, also for some coupled problems, which will be solved with so-called Deep Energy Methods (DEM).

3.2 Second-Order PDEs

3.2.1 *Elliptic PDEs*

3.2.1.1 Laplace, Poisson and Helmholtz Equation

The Laplace, Poisson and Helmholtz equations are popular elliptic linear PDEs in engineering. In fact, the Laplace equation is a special case of the Poisson equation, which in turn is a special case of the inhomogeneous Helmholtz equation that can be derived from the (hyperbolic) wave equation through separation of variables. The inhomogeneous Helmholtz equation can be written as

$$\Delta u(x_i) + k^2 u(x_i) = g(x_i) \tag{3.20}$$

where $\Delta = \nabla^2 = \nabla \cdot \nabla$ denotes the Laplace operator and k is the wave number, which can be related to the speed of sound c and the angular frequency $\omega = 2\pi f$ (f being the frequency), i.e. $k = \frac{\omega}{c}$; $g(x_i)$ is a source term (also called inhomogeneous term) and $u(x_i)$ is the primary variable. The RHS is zero for the homogeneous Helmholtz equation, which has applications in the study of various physical phenomena, such as acoustics, seismology and electromagnetic radiation. For many problems, the domain Ω is not bounded and the solution $u(x_i)$ can be highly oscillatory, which creates difficulties in most numerical methods. The solution becomes particularly challenging for high wave numbers k (usually in the order of thousands) and small wave lengths $\lambda = \frac{2\pi}{k}$ when the oscillatory behaviour becomes more pronounced. The Poisson equation does not contain the second term on the RHS of Eq. (3.20) leading to

$$\Delta u(x_i) = g(x_i) \tag{3.21}$$

while the source term vanishes for the Laplace equation

$$\Delta u(x_i) = 0. \tag{3.22}$$

For a unique solution, the PDE needs to be complemented with appropriate boundary conditions as mentioned in Sect. 3.1. One key characteristics of elliptic PDEs such as the Laplace equation is their smoothness. Indeed, information about the 'nature'—such as smoothness and regularity—of the solution is important for the machine learning based solution of PDEs such as setting up the network, choosing appropriate activation functions or transfer learning. This has also been the basis of extended finite element solutions which have introduced (local) enrichment functions in order to improve the accuracy of the numerical solution. For instance, re-entrant corners in the geometry or the presence of cracks affect the regularity of the solution of the Laplace or Poisson equation, i.e. they are governed by a singularity at the corner or the crack tip.

3.2.1.2 Linear Elasticity

The governing equation of the underlying problem is the equilibrium equation which can be derived from conservation of linear momentum postulating that the change in linear momentum is equal to the external forces acting on the system. Recall the definition of linear momentum, $\mathbf{p} = m\dot{\mathbf{x}}$ or in integral form:

$$\mathbf{p} = \int_{\Omega} \varrho \dot{\mathbf{x}} \, d\Omega \tag{3.23}$$

where ϱ denotes the density in the current configuration and the superimposed 'dot' indicates material time derivatives $(\dot{\square}) = \frac{d(\square)}{dt}$. The change in linear momentum is then defined accordingly by

$$\frac{d\mathbf{p}}{dt} = \frac{d}{dt}\left(\int_{\Omega} \varrho \dot{\mathbf{x}} \, d\Omega\right) \tag{3.24}$$

Taking advantage of Reynold's transport theorem, which states that the material time derivative of a domain integral over a specific quantity can be replaced by the integral of the material time derivative of this quantity, if and only if one integrated over a constant mass, i.e. $\frac{d}{dt}\left(\int_{\Omega} \varrho f \, d\Omega\right) = \int_{\Omega} \varrho \frac{df}{dt} \, d\Omega$, we obtain

$$\frac{d\mathbf{p}}{dt} = \int_{\Omega} \varrho \ddot{\mathbf{x}} \, d\Omega \tag{3.25}$$

The external forces acting on the system are the body forces and tractions acting along the Neumann boundary, so that conservation of linear momentum reads

$$\frac{d\mathbf{p}}{dt} = \int_{\Omega} \varrho \ddot{\mathbf{x}} \, d\Omega = \int_{\Omega} \varrho \mathbf{b} \, d\Omega + \int_{\Gamma_t} \bar{\mathbf{t}} \, d\Gamma \tag{3.26}$$

Using Cauchy theorem, i.e. the relation between the stress vector (the traction) and the stress tensor, leads to

$$\int_{\Omega} \varrho \ddot{\mathbf{x}} \, d\Omega = \int_{\Omega} \varrho \mathbf{b} \, d\Omega + \int_{\Gamma_t} \boldsymbol{\sigma} \cdot \mathbf{n} \, d\Gamma \tag{3.27}$$

Employing Gauss divergence theorem finally yields

$$\int_{\Omega} \varrho \ddot{\mathbf{x}} \, d\Omega = \int_{\Omega} \varrho \mathbf{b} \, d\Omega + \int_{\Omega} \nabla \cdot \boldsymbol{\sigma} \, d\Omega \tag{3.28}$$

The integrand of this expression is the well known equation of motion, which is a hyperbolic PDE. If the inertia term vanishes, we obtain the equilibrium equation (being elliptic):

$$\nabla \cdot \boldsymbol{\sigma} + \varrho \mathbf{b} = 0 \tag{3.29}$$

Note that the equilibrium equation could have been derived 'directly' in strong form by taking force equilibrium at an infinitesimal volume element. As already pointed out, the equilibrium equation is complemented by constitutive equations, kinematic equations and appropriate boundary conditions:

Constitutive equations :

$$\boldsymbol{\sigma} = \mathbb{C} : \boldsymbol{\epsilon}, \tag{3.30}$$

Kinematic equations :

$$\boldsymbol{\epsilon}(\boldsymbol{u}(\boldsymbol{x})) = \frac{1}{2}\left(\nabla \boldsymbol{u} + (\nabla \boldsymbol{u})^T\right) \tag{3.31}$$

Boundary conditions :

$$\mathbf{u} = \bar{\mathbf{u}} \text{ on } \Gamma_u \tag{3.32}$$

$$\boldsymbol{\sigma} \cdot \mathbf{n} = \mathbf{t} = \bar{\mathbf{t}} \text{ on } \Gamma_t \tag{3.33}$$

The boundary value problem of linear elastostatics plays a major role in the design of structures in civil and mechanical engineering. Its extension to geometrical and material nonlinearities has numerous other applications in various disciplines. Geometrical nonlinearities for instance are needed for phenomena such as buckling or large deformation analysis.

3.2.2 *Parabolic PDEs*

3.2.2.1 Convection-Diffusion and Diffusion Equation

The convection-diffusion equation describes mathematically transport phenomena through fluids or solids such as the transport of energy, heat, mass, charge, etc. Often the transport equations are solved together with other PDEs. For instance, the continuity equation which ensures conservation of mass is commonly solved with the conservation of linear momentum and conservation of energy in fluid and solid mechanics; note that for solids it is more common to use an algebraic equation to update the density ($\varrho_0 = J\varrho$, $J = det(\mathbf{F})$ being the determinant of the deformation gradient, which describes the volume change and ϱ_0 is the initial density). The energy equation is of particular interest in impact engineering and high dynamic loading rates with associated (at least locally) high hydrostatic pressures. In this case, constitutive models take advantage of splitting the Cauchy stress tensor into a deviatoric and hydrostatic part. The latter part commonly relates the hydrostatic pressure to the internal energy.

As the name already indicates, the convection-diffusion equation concerns two popular transport mechanisms, i.e. convection and diffusion. 'Convection typically refers to the transfer of heat due to the bulk movement of molecules within fluids' and encompasses both advection and diffusion. Advection denotes the transport of a quantity due to the motion of another quantity it is immersed in. This phenomenon is inherent to fluid flow and barely observed in solids. Diffusion is the transport of a quantity from a higher concentration to a lower concentration and the driving mechanism is the gradient of the associated quantity. It appears in both solids and fluids. A classical example for diffusion is the transport of energy due to temperature gradients, which is commonly referred to as heat conduction. Here, the flux is related to the temperature gradient through the thermal conductivity, which is also known as Fourier's law, see also Table 3.1 containing other examples. Heat can also be advected but in this case, the heat would usually 'travel' with the medium/fluid it transports it. Another diffusion driven transport mechanism is the transport of momentum, which is related to velocity gradients in a fluid. It is called viscous flow as it expresses the shear stress through the viscosity to the spatial velocity gradient.

The convection-diffusion equation is given by

$$\frac{\partial u}{\partial t} = \nabla \cdot (D \nabla u) - \nabla \cdot (\mathbf{v}\, u) \tag{3.34}$$

where u is the primary variable, t is the time, D the diffusivity determining the rate of diffusion and $\mathbf{v}$ the velocity u is moving with. If the primary variable u is the temperature (commonly indicated by T in this case), the diffusivity $D = \frac{k}{c_p \varrho}$ is called thermal diffusivity which is related to the thermal conductivity k in Table 3.1, the specific heat capacity c_p and the density ϱ. Note that we simplified assumed the thermal conductivity as scalar though it is in the most general (anisotropic) case a second-order tensor. If the diffusivity is constant and for incompressible flow, i.e. $\nabla \cdot v = 0$, the convection diffusion equation simplifies to

$$\frac{\partial u}{\partial t} = D \nabla \cdot \nabla u - \mathbf{v} \cdot \nabla u = D \Delta u - \mathbf{v} \cdot \nabla u \tag{3.35}$$

The first term on the RHS of Eq. (3.35) is diffusion-driven transport while the second term is advection-driven transport of quantity u. As already mentioned previously, advection is not present in solids. So, neglecting the last term on the RHS of Eq. (3.35) yields the diffusion equation—assuming a constant diffusivity:

$$\frac{\partial u}{\partial t} = D \Delta u \tag{3.36}$$

or in the more general case:

$$\frac{\partial u}{\partial t} = \nabla \cdot (D \nabla u) \tag{3.37}$$

As elliptic PDEs, parabolic PDEs such as the diffusion equation smooth out the initial data. Furthermore, they have an infinite propagation speed. In the stationary case, the diffusion equation is simplified to

$$\nabla \cdot (D\nabla u) = 0 \tag{3.38}$$

as already mentioned in the introduction Sect. 3.1 where we already presented some potential engineering applications. Note that the stationary diffusion equation is apparently an elliptic PDE while the time-dependent diffusion equation is parabolic, which in turn matches the title of this subsection. Let us come back to Eq. (3.35). There are a few applications where the diffusivity vanishes, for instance at temperature of absolute zero. In this case, the equation becomes

$$\frac{\partial u}{\partial t} + \mathbf{v} \cdot \nabla u = 0 \tag{3.39}$$

An important application of the advection equation in computational mechanics is in moving boundary problems when the boundary is implicitly described by a level set function. The advection equation needs to be also fulfilled by the level set function and is employed to update the level set function in applications in fluid mechanics, biofilm or tumor growth. In this context, level sets have been successfully combined with the extended finite element method (XFEM). They can also been used in fracture problems though their application is not straightforward as—in contrast to fluids—the propagation is *not* orthogonal to the boundary and the velocity $\mathbf{v}$ (the crack speed) in Eq. (3.39) is generally not known.

3.2.2.2 Navier-Stokes Equation

The Navier-Stokes equation can be derived from the linear momentum equation given in strong form by

$$\varrho \frac{d\mathbf{v}}{dt} = \nabla \cdot \boldsymbol{\sigma} + \varrho \mathbf{b} \tag{3.40}$$

$\mathbf{v}$ denoting the velocity. For applications in solid mechanics, a Lagrangian description of motion is commonly applied and the velocity is expressed as a function of the material coordinates $\mathbf{X}$ and the time t. Hence, the material time derivatives of the velocity field is identical to the partial time derivatives. However, in fluid mechanics the velocity is a function of the spatial coordinates $\mathbf{x}$ such that the material time derivatives contains also a convective term. Hence, the linear momentum equation reads

$$\varrho \frac{\partial \mathbf{v}}{\partial t} + \mathbf{v} \cdot \nabla \mathbf{v} = \nabla \cdot \boldsymbol{\sigma} + \varrho \mathbf{b} \tag{3.41}$$

Decomposing the Cauchy stress tensor $\boldsymbol{\sigma} = \boldsymbol{\tau} - p$ into a deviatoric $\boldsymbol{\tau}$ and hydrostatic part $p = -\frac{1}{3} tr\, \boldsymbol{\sigma}$ then leads to

$$\varrho \frac{\partial \mathbf{v}}{\partial t} + \mathbf{v} \cdot \nabla \mathbf{v} = -\nabla p + \nabla \cdot \boldsymbol{\tau} + \varrho \mathbf{b} \tag{3.42}$$

For a Newtonian (viscous) fluid, $\boldsymbol{\tau}$ is related to the spatial velocity gradient $\mathbf{l} = \frac{\partial \mathbf{v}}{\partial \mathbf{x}}$ through the fluid's viscosity ν $\boldsymbol{\tau} = \nu \left(\mathbf{l} + \mathbf{l}^T\right)$, which finally leads to the compressible Navier-Stokes equation

$$\varrho \frac{\partial \mathbf{v}}{\partial t} + \mathbf{v} \cdot \nabla \mathbf{v} = -\nabla p + \nabla \cdot \left[\nu \left(\mathbf{l} + \mathbf{l}^T\right)\right] + \varrho \mathbf{b} \tag{3.43}$$

For the incompressible Navier-Stokes equation, where the viscosity is constant, an additional 'constraint equation' has to be fulfilled, i.e.

$$\begin{aligned} \varrho \frac{\partial \mathbf{v}}{\partial t} + \mathbf{v} \cdot \nabla \mathbf{v} &= -\nabla p + \nu \nabla \cdot \left[\left(\mathbf{l} + \mathbf{l}^T\right)\right] + \varrho \mathbf{b} \\ \nabla \cdot \mathbf{v} &= 0 \end{aligned} \tag{3.44}$$

Note that the second equation (the constraint equation) can also be formulated in terms of the continuity equation, i.e. $\frac{\partial \varrho}{\partial t} + \nabla (\varrho \mathbf{v}) = 0$ which states that the material time derivatives of the density vanishes (no change in density). For inviscid fluids, when the viscosity is zero, the second term on the RHS of Eq. (3.44) is zero. Another special case is Stokes flow, if inertia forces are small and can be neglected such that the LHS of Eq. (3.44) is zero. This might be comparable to the equilibrium equation for solids.

3.2.3 Hyperbolic PDEs

3.2.3.1 Wave Equation

The understanding and research in (propagation of) waves has drastically changed our modern life. The applications of waves is a tremendous wide area affecting nearly all disciplines and industries including medicine, physics, engineering, etc. The entire wireless communication for instance is based on (electromagnetic) waves. Waves propagate through any medium and transport specific information through signals. There are numerous types of waves including gravitational waves, matter waves, quantum-mechanical waves, electromagnetic waves and mechanical waves. The governing equation of quantum-mechanical waves for instance is the Schrödinger equation, which plays a crucial role in quantum mechanics. It is also the basis of DFT simulations, that build an entire branch in materials science and are of utmost importance for e.g. the development of new materials. The physics of electromagnetic waves is mathematically described by Maxwell's equations, probably *the* most studied PDE in engineering and physics. Electromagnetic waves are caused by variations of magnetic and electric field intensities. Microwaves, infrared and ultraviolet waves, radio waves, X-ray and gamma rays all belong to the class of electromagnetic waves. They can also travel through a vacuum, in this case with a constant speed, i.e. the speed of light ($c = 3 \times 10^8$ m/s). Mechanical waves are oscillations of matter transferring energy through a medium. They can be classified into three types: In *longitudinal waves*, the displacement of the medium moves parallel to the travelling wave. Classical examples of longitudinal waves are sound waves, pressure waves or ultrasound waves. *Transverse waves* cause vibration of the medium, which are perpendicular to the traveling wave while *surface waves* travel along interfaces or surfaces and can be further categorized into Rayleigh and Love waves.

Particularly challenging is the numerical treatment of shock waves (see also the next subsection), which travel with supersonic speed and are characterized by a nearly discontinuous wave front. They occur in military applications, for instance when solids are subject to high dynamic loading as they occur in contact detonations (or generally explosive loading) or high-velocity impacts.

Waves are also of major importance in the area of photonics and phononics. Photon is the transportation of electromagnetic waves or microwaves, while phonons are quantizations of atomistic excitation in a lattice structure (or) quantization of wave propagation due to lattice distortion in a discrete system. When the long wave length limit is reached, it represents an elastic wave in a continuum such as a stress wave and acoustic wave. One extremely popular branch in these areas is the design of metamaterials. Quantum metamaterials for instance are expected to enable quantum computers to overcome current computing limits [15]. Topological insulators are probably the most exciting and fastest growing area crossing physics and materials science among other different types of metamaterials. In 2016, the Nobel Prize in physics was awarded on theoretical discoveries of topological phase transitions and topological phases of matter. Topological states have been discovered in electronic systems in the past decades. A topological phononic material offers a topological state with protected propagation of elastic waves. The topological indices of phononic metamaterials correspond to the vibration modes or elastic waves which also demonstrate state changes. Though topological insulators were originally developed in the area of optics and electronics, recently smart phononic and mechanical metamaterials such as free motions, load bearing states or vibrations, have emerged with analogues to topological protected states as the counterparts of photonic metamaterials. These materials can intelligently and adaptively respond to changes in the external thermal and mechanical stimuli simultaneously.

The wave equation is a linear second-order PDE which mathematically describes the propagation of waves as the name already suggests. It has applications in acoustics, electromagnetics, fluid and solid mechanics. It was first introduced in 1D by d'Alembert in 1752 and extended to 2D and 3D by Bernoulli 3 years later. In

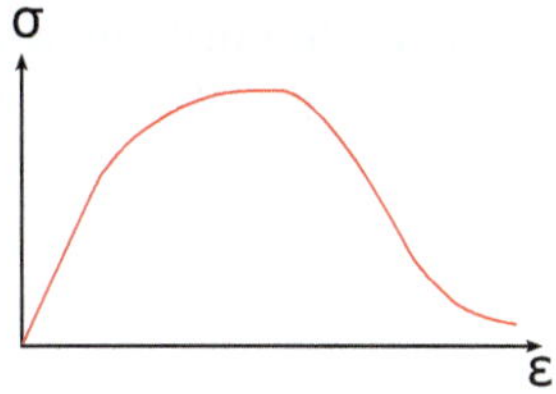

Fig. 3.3 Material exhibiting strain softening

higher order dimension, the wave equation is given by

$$\frac{\partial^2 u}{\partial t^2} = c^2 \Delta u \tag{3.45}$$

$c > 0$ denoting the wave speed. For a linear elastic solid, the wave speed can be related to the Young's modulus E and current density ϱ. For non-linear material behaviour, the Young's modulus E in Eq. (3.45) is to be replaced with the consistent material tangent stiffness. When the constitutive behaviour exhibits strain softening, i.e. the stresses decrease at some point with increasing strains as illustrated in Fig. 3.3, the tangent stiffness becomes negative and the wave speed imaginary, which physically means a standing wave. Mathematically, it indicates the transition from a hyperbolic to an elliptic PDE. At the special case, where the tangent stiffness is zero, i.e. the stress-strain curve reaches its maximum, the PDE is parabolic. Note also, that the initial data is not smoothed out for hyperbolic PDEs. In 1D, the equation simplifies to

$$\frac{\partial^2 u}{\partial t^2} = c^2 \frac{\partial^2 u}{\partial x^2} \tag{3.46}$$

which is the only wave equation that can be solved in closed form on an infinite domain. This solution is also known as d'Alembert solution.

3.2.3.2 Burgers' Equation and Linear Elastodynamics

Burgers' equation is a non-linear differential equation, which can be derived from the Navier Stokes equation by ignoring the pressure gradient [16]. It has been first introduced in 1915 by Harry Bateman before Johannes Martinus Burgers took advantage of his equation to mathematically explain turbulences in 1948. Eberhard Hopf and Julian Cole converted Burgers' equation into a linear heat equation 2 years later. For fluids with small viscosity, Burgers' equation leads to a shock formation, which is difficult to resolve by classical numerical methods. In 1D, Burgers' equation for the primary variable $u(x, t)$ is given by

$$\frac{\partial u}{\partial t} + u \frac{\partial u}{\partial x} = \nu \frac{\partial^2 u}{\partial x^2} \tag{3.47}$$

where ν refers to the dynamic viscosity. The second term on the LHS is the advective term while the RHS indicates the diffusive term. When $\nu = 0$, we obtain the inviscid Burgers's equation:

$$\frac{\partial u}{\partial t} + u \frac{\partial u}{\partial x} = 0 \tag{3.48}$$

The practical application of Burgers' equation is clearly in fluid mechanics such as gas dynamics, traffic and viscous flow, shallow water and shock waves, cosmology or turbulences, to name a few.

The governing equations of elastodynamics for a linear isotropic solid is a special case of Newton's second law given by

$$\frac{\partial^2 u_i}{\partial t} = \frac{\partial \sigma_{ij}}{\partial x_j} + \varrho b_i \tag{3.49}$$

Let us express the constitutive equations w.r.t. its Lame constants

$$\sigma_{ij} = \lambda \theta \delta_{ij} + 2\nu \epsilon_{ij} \tag{3.50}$$

where $\theta = tr(\boldsymbol{\varepsilon})$ is the dilatation. Substituting the kinematic equations $\epsilon_{ij} = 0.5(u_{i,j} + u_{j,i})$ into the constitutive equation and subsequently the Cauchy stress tensor back into the equation of motion yields

$$\varrho \frac{\partial^2 u_i}{\partial t^2} = \lambda \frac{\partial^2 u_j}{\partial x_i \partial x_j} + \nu \frac{\partial^2 u_j}{\partial x_i \partial x_j} + \nu \frac{\partial^2 u_i}{\partial x_j \partial x_j} + \varrho b_i \tag{3.51}$$

or in compact notation

$$\varrho \frac{\partial^2 \mathbf{u}}{\partial t^2} = (\lambda + \nu) \nabla \nabla \cdot \mathbf{u} + \nu \nabla^2 \mathbf{u} + \varrho \mathbf{b} \tag{3.52}$$

Substituting the well known relation $\nabla^2 \mathbf{u} = \nabla \nabla \cdot \mathbf{u} - \nabla \times \nabla \times \mathbf{u}$ into above equation finally results in

$$\varrho \frac{\partial^2 \mathbf{u}}{\partial t^2} = (\lambda + 2\nu) \nabla \nabla \cdot \mathbf{u} + \nu \nabla \times \nabla \times \mathbf{u} + \varrho \mathbf{b} \tag{3.53}$$

which is also known as Navier equation.

3.3 Higher-Order PDEs

In this section, we will briefly present two higher-order PDEs which are well-known to any undergraduate student in engineering, i.e. the Euler-Bernoulli beam and the Kirchhoff plate. They are the basis for the design of structures in components.

3.3.1 *Euler-Bernoulli Beam*

Let us consider a straight beam made of a homogeneous isotropic material. The beam has a length L and the cross-sectional dimensions are assumed to be significantly smaller compared to L such that transverse shear effects can be neglected. The deflection w of the beam only depends on x and plane cross sections remain plane. Furthermore, we assume small strain theory, such that the displacement $u(x, y)$ for an arbitrary point in the cross-section is obtained according to Fig. 3.4 by

$$u(x, y) = z\beta(x) \tag{3.54}$$

The Euler-Bernoulli constraint ensures zero shear strains, such that the angle β can be expressed in terms of the first partial derivative of the deflection w, i.e. $\frac{\partial w}{\partial x} = -\beta(x)$. Since w depends only on x, the only non-zero strain component is the normal strain in x-direction, i.e.

$$\epsilon_{xx} = \frac{\partial u}{\partial x} = z\frac{\partial \beta(x)}{\partial x} = -z\frac{\partial^2 w(x)}{\partial x^2} \tag{3.55}$$

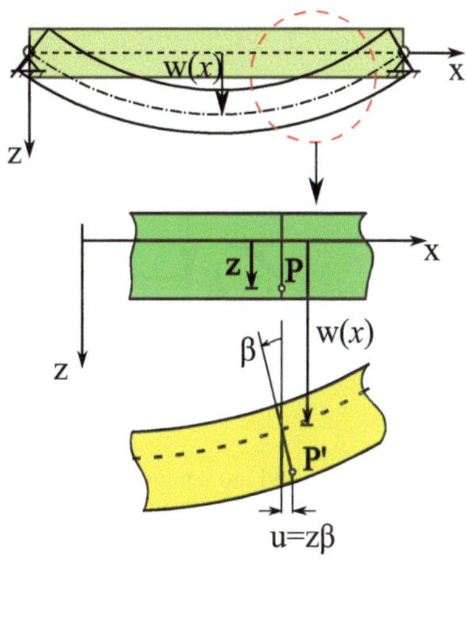

Fig. 3.4 Kinematics of the Euler Bernoulli beam

The efficiency of structural elements such as beams, plates or shells are related to the dimension reduction. Therefore, instead of working with stresses and strains, stress and strain resultants are used. The relevant strain resultant is the curvature, which is nearly equal to the (negative) second derivative of w for small strain theory, i.e. $\kappa(x) \approx -\frac{\partial^2 w(x)}{\partial x^2}$. The bending moment $M_y(x)$, i.e. the relevant stress resultant for Euler Bernoulli beams, can be related to the normal stress $\sigma_{xx}(x, z)$ through

$$M_y = \int_A \sigma_{xx} z \, dA \tag{3.56}$$

A denoting the cross-sectional area. Substituting Eq. (3.55) into Hooke's law $\sigma_{xx} = E\,\epsilon_{xx}$ (E being the elasticity modulus) and subsequently into Eq. (3.56) yields

$$M_y = -E \frac{\partial^2 w(x)}{\partial x^2} \int_A z^2 \, dA \tag{3.57}$$

Defining the second moment of inertia $I_y = \int_A z^2 \, dA$, the well known $M - \kappa$ relation is obtained

$$M_y = -E I_y \frac{\partial^2 w(x)}{\partial x^2} \tag{3.58}$$

where EI_y indicates the bending rigidity. Assume now that the beam is loaded only by a distributed load $q_z(x)$ normal to the mid-line of the beam and consider the force (in z-direction) and moment equilibrium for an infinitesimal element, as illustrated in Fig. 3.5. This leads to

$$\begin{aligned} \frac{dQ_z}{dx} &= q_z(x) \\ dM_y - \frac{q_z\, dx^2}{2} &= Q_z \, dx \end{aligned} \tag{3.59}$$

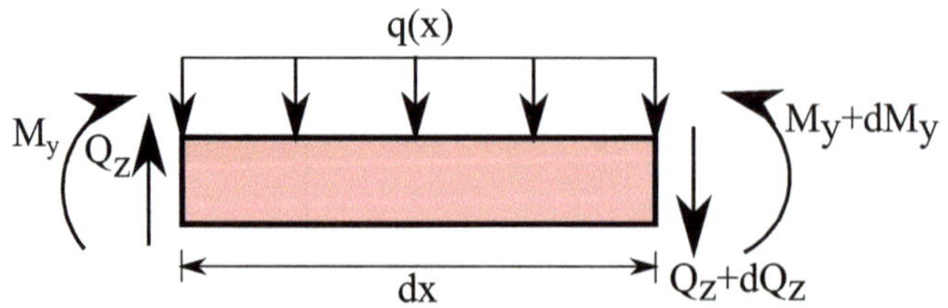

Fig. 3.5 Force and moment equilibrium for an infinitesimal element of the Euler Bernoulli beam

The second term of the RHS is neglected yielding

$$\frac{dQ_z}{dx} = q_z(x)$$
$$\frac{dM_y}{dx} = Q_z \tag{3.60}$$

and hence $\frac{d^2M_y}{dx^2} = -q_z$, which is the governing equation, which can be expressed in terms of the primary variable, i.e. the deflection of the beam $w(x)$ by

$$EI_y \frac{\partial^4 w(x)}{\partial x^4} = q_z \tag{3.61}$$

which is obviously a fourth-order differential equation, that needs to be complemented with appropriate boundary conditions, i.e.

$$\frac{dM_y}{dx} = Q_z = \bar{Q} \text{ on } \Gamma_t$$
$$M_y = \bar{M} \text{ on } \Gamma_t$$
$$u = \bar{u} \text{ on } \Gamma_u$$
$$w = \bar{w} \text{ on } \Gamma_u$$
$$\frac{\partial w}{\partial x} = \bar{\beta} \text{ on } \Gamma_u \tag{3.62}$$

Γ_t and Γ_u being the traction and displacement boundaries and superimposed 'bars' indicating imposed (boundary) values. Note that the axial component has been neglected though such an extension can be easily added.

3.3.2 *Kirchhoff Plate*

Kirchhoff plate theory has been proposed in 1850 by Gustav Robert Kirchhoff and is a 'straight-forward' extension of the Euler-Bernoulli beam theory. Assuming again a linear elastic isotropic material, small strain theory and that the deflection w depends only on x and y, we can write the displacement components of a material point in the cross-section according to Fig. 3.6:

$$u(x, y, z) = z\beta_x$$
$$v(x, y, z) = z\beta_y \tag{3.63}$$

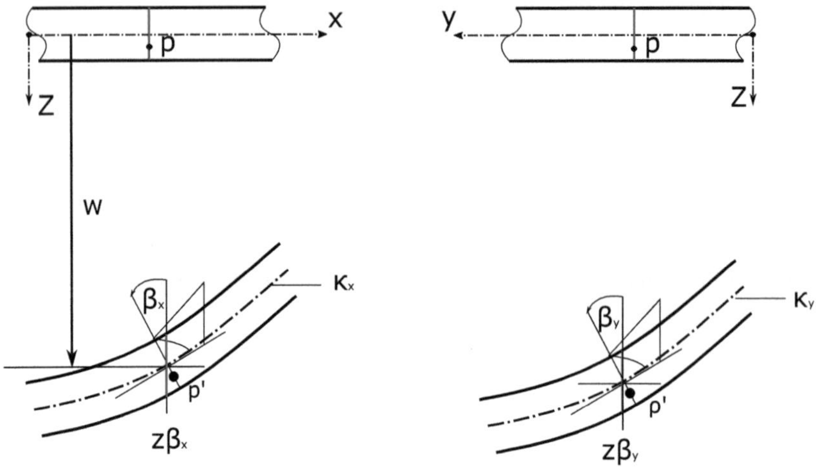

Fig. 3.6 Kinematics for the Kirchhoff plate

where u and v refer to the displacements in x- and y-direction, respectively. Note that β_x is defined positive rotating clockwise while β_y is positive for a counter-clockwise rotation. We assume again that the normal to the midsurface position remains inextensible and normal to the midsurface in the deformed configurations. Furthermore, plane cross sections remain plane. One first notices that the transverse shear strains γ_{xz} and γ_{yz} vanish and similarly to the Euler-Bernoulli theory, there exists a relation between the derivatives of the deflection w and the angles β_x and β_y:

$$\begin{aligned} \frac{\partial w}{\partial x} &= \beta_y \\ \frac{\partial w}{\partial y} &= \beta_x \end{aligned} \tag{3.64}$$

The curvatures can be approximated as in Euler-Bernoulli theory by

$$\begin{aligned} -\frac{\partial^2 w}{\partial x^2} &\approx \kappa_y \\ -\frac{\partial^2 w}{\partial y^2} &\approx \kappa_x \\ -\frac{\partial^2 w}{\partial x \partial y} &\approx \kappa_{xy} \end{aligned} \tag{3.65}$$

where the last term denotes the twisting, a term not present in Euler-Bernoulli theory. The associated components of the strain tensor are now obtained by

$$
\begin{aligned}
\epsilon_{xx} &= z\kappa_y \\
\epsilon_{yy} &= z\kappa_x \\
\gamma_{xy} &= z\kappa_{xy}
\end{aligned}
\tag{3.66}
$$

Substituting Eq. (3.66) into the constitutive model given by

$$
\begin{aligned}
\sigma_{xx} &= \frac{E}{1-\nu^2}\left(\epsilon_{xx} + \nu\epsilon_{yy}\right) \\
\sigma_{yy} &= \frac{E}{1-\nu^2}\left(\epsilon_{yy} + \nu\epsilon_{x}\right) \\
\tau_{xy} &= G\gamma_{xy}
\end{aligned}
\tag{3.67}
$$

and the outcome into

$$
\begin{aligned}
m_x &= \int_{-t/2}^{t/2} z\sigma_{xx}\, dz \\
m_y &= \int_{-t/2}^{t/2} z\sigma_{yy}\, dz \\
m_{xy} &= \int_{-t/2}^{t/2} z\tau_{xy}\, dz
\end{aligned}
\tag{3.68}
$$

yields

$$
\begin{aligned}
m_x &= \frac{E}{1-\nu^2}\int_{-t/2}^{t/2} z^2\left(\kappa_x + \nu\kappa_y\right)\, dz = \frac{E\, t^3}{1-\nu^2}\left(\kappa_x + \nu\kappa_y\right) \\
m_y &= \frac{E}{1-\nu^2}\int_{-t/2}^{t/2} z^2\left(\kappa_y + \nu\kappa_x\right)\, dz = \frac{E\, t^3}{1-\nu^2}\left(\kappa_y + \nu\kappa_x\right) \\
m_{xy} &= \frac{G\, t^3\, \kappa_{xy}}{6} = \frac{E\, t^3}{12(1+\nu)}\kappa_{xy}
\end{aligned}
\tag{3.69}
$$

where t designates the plate's thickness and $\frac{E\, t^3}{12(1-\nu^2)}$ the bending stiffness, $G = \frac{E}{2(1+\nu)}$ is the shear modulus. Assuming only loads perpendicular to the plate surface $q_z(x, y)$, force equilibrium (in z-direction) as well as moment equilibrium around the x- and y-axis leads to

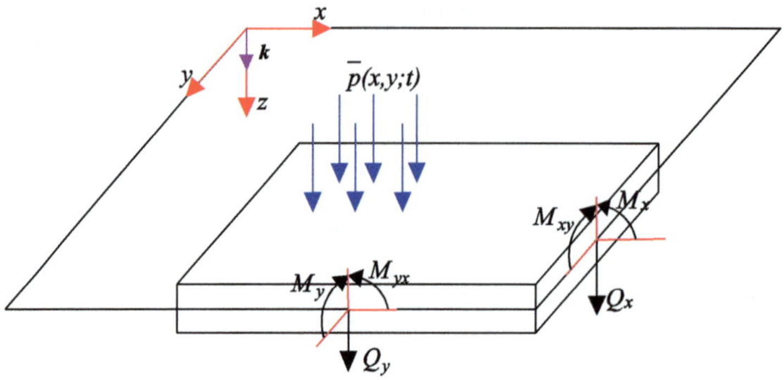

Fig. 3.7 Kirchhoff plate

$$\frac{\partial Q_x}{\partial x} + \frac{\partial Q_y}{\partial y} + q_z = 0$$
$$\frac{\partial m_x}{\partial x} + \frac{\partial m_{xy}}{\partial y} - Q_x = 0$$
$$\frac{\partial m_y}{\partial y} + \frac{\partial m_{xy}}{\partial x} - Q_y = 0 \tag{3.70}$$

Similarly to the Euler-Bernoulli beam, the shear forces Q_x and Q_y are related to the bending moments m_x and m_y (see Fig. 3.7). Hence, solving the last line of Eq. (3.70) w.r.t. Q_y and the line before w.r.t. Q_x and substituting the outcome into the first line of Eq. (3.70) results in the final governing equation of the Kirchhoff plate:

$$\frac{\partial^2 m_x}{\partial x^2} + \frac{\partial^2 m_y}{\partial y^2} + 2\frac{\partial m_{xy}}{\partial x \partial y} + \bar{p}(x, y; t) = 0 \tag{3.71}$$

Substituting the constitutive equation, i.e. the $m - \kappa$ relations, into the governing equation results in the biharmonic equation:

$$\frac{E\,t^3}{12(1 - \nu^2)} \Delta\Delta w - \bar{p}(x, y; t) = 0 \tag{3.72}$$

These equations need to be complemented by appropriate boundary conditions.

3.4 Energy Methods

Many engineering problems can be expressed in terms of an energy (or energy density), which—together with the principle of the minimum of the total potential energy—builds a natural basis for a machine learning based solution of BVPs.

In contrast to FEM, which finally results in a solution of an algebraic system of equations, the machine learning based solution of a BVP leads to a—commonly nonconvex—optimization problem. Note that the discrete system of equations in FEM can also be derived from the principle of the minimum of the total potential energy. As briefly mentioned in the introduction, concepts such as entropy and convexity play a significant role in the solution of PDEs, which are in turn major concepts in energy approaches. One classical application is constitutive modeling, where the existence of a suitable energy density can guarantee major principles (in constitutive modeling or continuum mechanics in general) such as objectivity (= frame invariance), thermodynamic consistency, well-posedness and existence of solutions. Therefore, the energy density has to fulfill certain conditions. Convexity, or more detailed polyconvexity of the energy density is one of them and a requirement for the existence of solutions for example while entropy is a major ingredient in thermodynamics. The second law of thermodynamics, the Clausius Duhem inequality, e.g. states that the entropy rate has to be larger equal than zero; the latter case (equal zero) indicates a reversible process while the first condition (larger zero) an irreversible one. Constitutive models, which cannot fulfill the second law of thermodynamics, might lead to unphysical behaviour. More details about constitutive models can be found in nearly any continuum mechanics book [17, 18] while the importance of polyconvexity is discussed for instance in [19, 20].

3.4.1 Hyperelasticity

Hyperelastic—sometimes also called Green elastic—materials belong to the class of Cauchy elastic materials. In such materials, the stresses and strains are reversible and path-independent. In contrast to Cauchy elastic materials, hyperelastic materials are derived from a strain energy density function and therefore fulfill the basic equations of thermodynamics (other than Cauchy elastic materials). There are many materials including biological materials, rubber, certain polymers, etc. that can be well described through hyperelastic constitutive models. One advantage of hyperelastic constitutive models is the ease in implementation in finite element codes, also thanks to tools such as Acegen [21, 22] or Fenics [23] which are capable of automatic code generation. They require—more or less—only the definition of the strain energy density and provide the associated FE code. On the other hand, constitutive models in rate form, which are not based on an energy density, lead to issues such as objectivity due to the integration in time, even when objective rates such as the Jaumann-, Truesdell-, Green-Naghdi-, or (upper or lower) Oldroyd rate are used. This issue of incremental objectivity has been extensively discussed in the literature, see e.g. the seminal work of Hughes and Winget [24], Simo et al. [25–27] or Auricchio and Taylor [28]. One difficulty in hyperelastic constitutive models is the calibration of the material parameters, see e.g. the contribution of Gasser and Holzapfel [29] for biological materials. However, machine learning is also a perfect solution to this problem (calibration of material parameters). Associated tools have meanwhile already been coupled to commercial software such as ABAQUS [30]. Subsequently, we will present a brief introduction to hyperelastic models, i.e. isotropic homogeneous solids for purely mechanical problems. An extensive overview of hyperelastic models can be found in above mentioned references.

The starting point of hyperelasticity is a strain energy density Ψ, which is a function of some deformation/strain measure such as the deformation gradient $\mathbf{F}$, right Cauchy Green tensor $\mathbf{C} = \mathbf{F}^T \cdot \mathbf{F}$ or Green Lagrange-strain tensor $\mathbf{E} = 0.5\,(\mathbf{C} - \mathbf{I})$, $\mathbf{I}$ being the second-order identity tensor. Taking derivatives of the strain energy density w.r.t. $\mathbf{F}$ or $\mathbf{C}$ or $\mathbf{E}$, respectively, results in the thermodynamic conjugated stress measure. Deriving the equations for conservation of energy shows that the work conjugated quantity of the Cauchy-stress tensor is the rate of deformation tensor $\mathbf{D}$, the symmetric part of the (spatial) velocity gradient $\mathbf{L} = \dot{\mathbf{F}} \cdot \mathbf{F}^T$. Taking advantage of the Piola transformations, one can show that the 1. Piola Kirchhoff tensor (1. PK) is work conjugated to the deformation gradient (more precisely $\dot{\mathbf{F}}$) while the second Piola Kirchhoff tensor (2. PK) $\mathbf{S}$ is work conjugated to $\dot{\mathbf{E}} = 0.5\dot{\mathbf{C}}$. This can be expressed mathematically

$$\mathbf{P} = \frac{\partial \Psi(\mathbf{F})}{\partial \mathbf{F}} \tag{3.73}$$

The simplest hyperelastic model is a St. Venant-Kirchhoff material, in which the strain energy density is expressed in terms of the Green-Lagrange tensor as follows:

$$\Psi(\mathbf{E}) = 0.5\lambda\,(tr\;\mathbf{E})^2 + \nu\mathbf{E} : \mathbf{E} \tag{3.74}$$

λ and ν designating the two Lame constants. Taking derivatives of Ψ w.r.t. $\mathbf{E}$ leads to

$$\mathbf{S} = \frac{\partial \Psi(\mathbf{E})}{\partial \mathbf{E}} = \lambda\,(tr\;\mathbf{E})\,\mathbf{I} + 2\nu\mathbf{E} \tag{3.75}$$

which can also be expressed by

$$\mathbf{S} = \mathbb{C}^S : \mathbf{E} \tag{3.76}$$

with $\mathbb{C}^S = \lambda\mathbf{I} \otimes \mathbf{I} + \nu\,(\mathbf{I} \otimes \mathbf{I} + \mathbf{I} \otimes \mathbf{I})$. Note that this constitutive equation is similar to the Hooke's law in linear elasticity for small strain theory with the only difference that the Cauchy stress tensor is replace by the 2. PK and the linear strain tensor with the Green-Lagrange tensor, respectively. Such a material model is unfortunately not of much practical use outside small strain theory.

As mentioned earlier, we restrict our focus on homogeneous isotropic solids, in which case the strain energy density and the Green-Lagrange tensor or right Cauchy-Green tensor do not depend on the chosen material axis. Hence, the strain energy density can be expressed in terms of the invariants of **C**. For a Hadamard material, the strain energy density can be expressed by

$$\Psi(\mathbf{C}) = a\,(I-3) + b\,(II-3) + h(J) \tag{3.77}$$

where $I = tr\ \mathbf{C}$ and $II = 0.5\left([tr\ \mathbf{C}]^2 - tr\ \mathbf{C}^2\right)$ refer to the first and second invariants of the right Cauchy Green tensor while $h(J)$ is a convex function depending on the determinant of the deformation gradient $J = det\ \mathbf{F}$. Taking derivatives of Ψ w.r.t. **C** leads to the following constitutive model

$$\mathbf{S} = 2\left[a\mathbf{I} + \left(bII + 0.5\tilde{h}(J)\ J\right)\mathbf{C}^{-1} - bJ^2\mathbf{C}^{-2}\right] \tag{3.78}$$

For an incompressible material, the constraint $J = 1$ needs to be added. Neglecting the last term on the RHS of Eq. (3.77) leads to a Mooney-Rivlin material given by

$$\Psi(\mathbf{C}) = a\,(I-3) + b\,(II-3) \tag{3.79}$$

3.4.2 Coupled Problems

A thermodynamic consistent formulation requires the fulfillment of some balance equations, i.e. conservation of mass, conservation of linear and angular momentum and conservation of energy. These equations are complemented with another inequality, the so-called Clausius-Duhem inequality, which provides information about the reversibility (or irreversibility) of the process. Therefore, another quantity is required, the so-called entropy, which quantifies the irreversibility of the process. The Clausis-Duhem inequality states that the entropy rate has to be larger equal to zero.

3.4.2.1 Piezo/Flexoelectricity

Piezoelectricity represents electricity due to pressure and is a property of certain crystalline materials. The structure of such crystalline materials have a negative and positive polarization that neutralizes along the polar axis. When an external mechanical stress disturbs this charge balance, electric charge carriers create a current in the crystal. This is called the direct effect. On the other hand, a mechanical stress is created when an external charge creates an imbalance in the neutral charge state, which is termed as the converse effect. Piezoelectric materials under converse effect can be used as an actuator and under direct effect can be used as a sensor or energy transducer. The piezoelectric effect is attained by subjecting the crystalline material to a strong electric field at higher temperature. Materials possessing piezoelectric effects when deformed along the poling direction produce high voltage of the same or opposite polarity as that of the poling voltage, depending on the load direction. When a voltage difference is applied across the electrodes placed above and below the piezoelectric material, deformation takes place. Hence, they are commonly employed as energy harvesters transforming mechanical energy into electrical energy. An input AC signal leads to the vibration of the piezolayer at the same frequency as the AC signal. Commonly available piezoelectric materials are Lead Zirconium Titanate (PZT 4, PZT 5A, PZT 5H, PZT-8), Barium Titanate (BaTio3), Lead Magnesium Niobate Titanate (PMNPT), Polyvinylidene Fluoride (PVDF) and micro fibre composites (MFC).

Recently, enhanced electromechanical coupling at nanoscale has been discovered which was attributed to **flexoelectricity**, i.e. **the coupling between polarization and strain gradients**. The phenomenon of flexoelectricity was first detected by Mashkevich and Tolpygo [31] in the early 1950s but has received little attention, probably since flexoelectricity is relatively insignificant for bulk materials. However, the rapid advances in nano- and manufacturing technology in the past decades reactivated the interests on flexoelectric materials and their impact in many industrial sectors is promising. Most piezoelectric materials are brittle at the macro-scale and fail under small strains and small strain gradients. However, the vast technological progress in nanotechnology have allowed piezoelectric materials being manufactured at the nano-scale where the material can sustain much larger strain gradients, so that the influence of flexoelectricity becomes important. Studies revealed a drastic increase in voltage with decreasing length scale. This observation cannot be attributed to piezoelectricity alone. It was found in [32] for instance, that decreasing the thickness from $3 \cdot 10^{-6}$ m to $0.33 \cdot 10^{-6}$ m increases the electromechanical coupling coefficient by two orders of magnitude, which is attributed to flexoelectricity. Driven by the growing needs in developing compact self-powered devices such as portable electronics, wireless sensors and biomedical implants, flexoelectricity has recently attracted a lot of research interests.

Flexoelectricity is related to piezoelectricity, however it offers many additional electromechanical functions which are not provided by the latter. Firstly, flexoelectricity is a universal effect in all type of crystals (32 classes) and polymers while piezoelectricity is confined to non centrosymmetric crystals (20 classes). Therefore, it broadens the choice of materials for electromechanical devices. Secondly, flexoelectricity exhibits a strong size effect which develops quickly when the size goes below microscale as compared to piezoelectricity. As differences in the strain over a small distance (nanometers) yield large gradients and with nanomaterials that can survive much higher magnitudes of strain under buckling and wrinkling, flexoelectricity in nanoscale becomes significant. Flexoelectricity is highly sensitive to force and generates high strain gradients under the force of nano or even pico Newton. Finally, flexoelectricity can also be generated from crystals free of lead [33], which is the basic ingredient of ceramic piezoelectric materials. Lead free materials (nontoxic) are most appealing in biomedical application.

The combination of piezoelectricity and flexoelectricity leads to effective piezoelectricity which is polarity dependent. This asymmetry is obviously size dependent as the flexoelectric effect either works together or acts against the piezo-electric effect which in turn depends on the piezoelectric poling direction. This constructive and destructive interplay has led to an idea of electromechanical strain diodes [33]. The interplay between flexoelectricity and piezoelectricity can be exploited depending on the application required. For a bimorph actuator and sensor [34], different circuit conditions can lead to both either constructive and destructive interplay, which are size dependent.

Flexoelectric materials found so far are temperature dependent and usually peaks closer to Curie temperature. It has been tested that flexoelectric coefficients of barium titanate (BTO), barium strontium titanate (BST) and strontium titanate (STO) vary with temperature [35]. For BST thin films, μ_{12} of $17.44\,\mu C/m$ and $24.5\,\mu C/m$ were measured at 41 °C and 28 °C respectively [36]. In [37], the effective flexoelectric coefficient of BTO increases with temperature, and when above 250 °C it remains constant. Normally, flexoelectric coefficients are high for ferroelectrics with high dielectric constants. However, the dielectric constants of ferroelectrics decreases with increasing temperature, which is just the *opposite* for flexoelectricity. Narvaez [37] explained the reason as due to the free charge carriers arising from the surface piezoelectricity, which develops quickly after single crystal materials loose their insulating property at a temperature above 200 °C. Therefore, increased flexoelectricity can even occur above Curie temperature, which is not the case for piezoelectricity.

The contribution of surface effects and dynamic flexoelectricity in the total energy contribution is still not well understood. Experiments performed over a finite sample involve four possible contributions to the flexoelectric responses: static bulk flexoelectricity, dynamic bulk flexoelectricity, surface piezoelectricity and controversially surface flexoelectricity. In [38], it is claimed that the hundreds of volts obtained in experiments for BTO and STO, which clearly exceeds Kogan's estimate, cannot be only because of the static flexoelectric effect. The first contribution, static bulk flexoelectricity can be characterized by means of a fourth order material property tensor, called the bulk flexoelectric tensor [39] comparable in its form to the elastic tensor of materials. The second part, dynamic flexoelectricity, can be quantified by the flexodynamic tensor [40]. Dynamic flexoelectricity is the polarization due to the acceleration of the medium, and can be written in terms of strain gradients assuming the acceleration is negligible. Kvason et al. [40] extracted flexodynamic coefficients for cubic STO. The third part, surface piezoelectricity, comes from the surface relaxation leading to breaking of local symmetry and hence the charges appear along the surface. Some experimentalists estimate such effects can be even up to 100 nm deep from the surface and thereby justifying an increase in experimental tested coefficients. The existence of surface flexoelectricity remains still under debate. Resta et al. [41] has concluded that in the simplest class of dielectrics, there is "no surface contribution to flexoelectricity", and that it is therefore a pure bulk effect. Yudin et al. [39] on the other hand stated that the surface flexoelectric effect exists due to the non zero quadruple moment tensor. However, they also concluded that this effect is negligible for ferroelectrics with high dielectric constants and it seems that the surface flexoelectric effect is not of practical importance.

3.4.2.2 Formulation Based on the Internal Energy

For small strain theory, the internal energy for flexoelectricity can be expressed in terms of the linear strain tensor ϵ_{ij}, the strain gradient $\epsilon_{ij,k}$, polarization P_i and polarization gradient $P_{i,j}$ as

$$U\left(\epsilon_{ij}, \epsilon_{ij,k}, P_i, P_{i,j}\right) = \frac{1}{2}a_{ij}P_iP_j + \frac{1}{2}b_{ijkl}P_{i,j}P_{k,l} + \frac{1}{2}c_{ijkl}\varepsilon_{ij}\varepsilon_{kl} + d_{ijk}P_i\varepsilon_{jk} + f_{ijkl}\varepsilon_{ij}P_{k,l} + h_{ijkl}P_i\varepsilon_{jk,l} + \frac{1}{2}g_{ijklmn}\varepsilon_{ij,k}\varepsilon_{lm,n}. \tag{3.80}$$

For piezoelectricity, all terms on the RHS related to gradients vanish. The constitutive equations are obtained by taking the derivatives of the internal energy w.r.t. their primary variables yielding

$$\sigma_{ij} = \frac{\partial U}{\partial \varepsilon_{ij}} = c_{ijkl}\varepsilon_{kl} + d_{ijk}P_k + f_{ijkl}P_{k,l} \tag{3.81a}$$

$$\tau_{ijk} = \frac{\partial U}{\partial \varepsilon_{ij,k}} = h_{ijkl}P_l + g_{ijklmn}\varepsilon_{lm,n} \tag{3.81b}$$

$$E_i = \frac{\partial U}{\partial P_i} = a_{ij}P_j + d_{ijk}\varepsilon_{jk} + h_{ijkl}\varepsilon_{jk,l} \tag{3.81c}$$

$$V_{ij} = \frac{\partial U}{\partial P_{i,j}} = b_{ijkl}P_{k,l} + f_{ijkl}\varepsilon_{kl}, \tag{3.81d}$$

σ_{ij}, E_i, τ_{ijk} and V_{ij} denoting the (local) stress, electric field, higher-order stress and higher-order electric field, respectively and a_{ij}, b_{ijkl}, c_{ijkl}, d_{ijk}, f_{ijkl}, h_{ijkl} are the associated material tensors. Taking advantage of a variational principle proposed before by Kuang [42–45], Toupin [46] derived the governing equations for flexoelectricity accounting for the Maxwell's stress. We present the formulation suggest by Hu and Shen [47], which is an extension of Toupin's variational principle. Hence, the variation of internal energy can be written as

$$\delta U = \sigma_{ij}\delta\varepsilon_{ij} + \tau_{ijk}\delta\varepsilon_{ij,k} + E_i\delta P_i + V_{ij}\delta P_{i,j}. \tag{3.82}$$

The energy density is the sum of the internal energy the energy of the Maxwell self-field [48]

$$W = U + \frac{1}{2}\varepsilon_0\phi_{,i}\phi_{,i}, \tag{3.83}$$

in which the electric Maxwell self-field can be defined as the gradient of the electric potential ϕ

$$E_i^{MS} = -\phi_{,i}. \tag{3.84}$$

The electric enthalpy can be defined as

$$H = W - E_i D_i = U - \frac{1}{2}\varepsilon_0\phi_{,i}\phi_{,i} + \phi_{,i}P_i \tag{3.85}$$

Now, let us consider a dielectric solid occupying a volume v bounded by a surface a that separates from the surrounding environment v'; subsequently assumed to be air. For isothermal processes and defining $v^* = v + v'$, equilibrium can be written in variational form by

$$\delta \int_{v^*} H\,\mathrm{d}v^* = \delta W, \tag{3.86}$$

W referring to the external work due to the body force f_i, external electric field E_i^{ext}, free charge ρ^f on the solid body and traction $\bar{t}_i$, double-traction $\bar{\tau}_l$, surface charge σ^* and higher-order electric field $\bar{v}_i$ on the surface boundary leading to

$$\begin{aligned}\delta W = &\int_v \left(f_i\delta u_i + E_i^{\text{ext}}\delta P_i\right)\mathrm{d}v + \int_{a^\sigma} t_i\delta u_i\,\mathrm{d}a + \int_{a^\tau}\bar{\tau}_i\tilde{D}^n(\delta u_i)\,\mathrm{d}a\\ &+ \int_{a^V}\bar{v}_i\delta P_i\,\mathrm{d}a - \int_{a^\phi}\sigma^*\delta\phi\,\mathrm{d}a - \int_v \rho^f\delta\phi\,\mathrm{d}v,\end{aligned} \tag{3.87}$$

where $\tilde{D}^n = n_l\partial_{,l}(\circ)$ is the normal surface differential operator emerging from the treatment of the variation of the displacement gradient $\delta u_{i,j}$ on the boundary, see [47, 49] for more details. Kuang [42–45] showed that the variation of the electric enthalpy can be further expressed as

$$\delta\int_{v^*} H\,\mathrm{d}v^* = \int_{v^*}\delta H\,\mathrm{d}v^* + \int_{v^*} H^e\delta u_{k,k}\,\mathrm{d}v^*, \tag{3.88}$$

with $H^e = \frac{1}{2}\left(E_i P_i + V_{ij} P_{i,j}\right)$ being the energy deducted from the total energy minus the pure deformation energy. The migratory variation is the virtual displacement, that does not only produce the strain variation but also causes an electric potential and polarization variations yielding [47]

$$\delta\phi = \delta_\phi\phi + \delta_u\phi = \delta_\phi\phi + \phi_{,i}\delta u_i, \tag{3.89a}$$

$$\delta\phi_{,i} = \delta_\phi\delta\phi_{,i} + \phi_{,ij}\delta u_j + \phi_{,j}\delta u_{j,i} \tag{3.89b}$$

$$\delta P_i = \delta_P P_i + \delta_u P_i = \delta_P P_i + P_{i,j}\delta u_j, \tag{3.89c}$$

$$\delta P_{i,j} = \delta_P P_{i,j} + \delta_u P_{i,j} = \delta_P P_{i,j} + P_{i,jk}\delta u_k = \delta_P P_{i,j} + P_{i,kj}\delta u_k, \tag{3.89d}$$

where $\delta_\phi(\circ)$ and $\delta_P(\circ)$ are the variations produced by the (local) variation of the electric potential and polarization, respectively, while $\delta_u(\circ)$ is the migratory variation caused by the virtual displacement. Substituting Eq. (3.89) into Eq. (3.88) employing Eqs. (3.85) and (3.82) together with the Gauss divergence theorem, it can be shown that the variation of the electric enthalpy is given by [47]:

$$\begin{aligned}
\delta\int_v H\,\mathrm{d}v = &\int_v \left[-\sigma_{ij,j} + \tau_{ijm,mj} - \sigma^{ES}_{ij,j}\right]\delta u_i\,\mathrm{d}v \\
&+ \int_a \left\{\left[\sigma_{ij} - \tau_{ijm,m} + \sigma^{ES}_{ij}\right]n_j + \left[\tilde{D}^t_l(n_l)n_m n_j\tau_{ijm} - \tilde{D}^t_j\left(n_m\tau_{ijm}\right)\right]\right\}\delta u_i\,\mathrm{d}a \\
&+ \int_a \tau_{ijm}n_m n_j\tilde{D}^n(\delta u_i)\,\mathrm{d}a \\
&+ \int_v \left(E_i - V_{ij,j} + \phi_{,i} + E^{\mathrm{ext}}_i\right)\delta P_i\,\mathrm{d}v + \int_a V_{ij}n_j\delta P_i\,\mathrm{d}a \\
&- \int_v \left(-\varepsilon_0\phi_{,i} + P_i\right)_{,i}\delta\phi\,\mathrm{d}v + \int_a \left(-\varepsilon_0\phi_{,i} + P_i\right)n_i\delta\phi\,\mathrm{d}a\,,
\end{aligned} \tag{3.90}$$

n_i indicating the normal vector and $\tilde{D}^t_k(\circ) = (\delta_{kl} - n_k n_l)\,\partial_{,l}(\circ)$ the tangential surface differential operator, respectively. As a direct result of the variational process, the generalized electrostatic stress σ^{ES}_{ij} appears as

$$\begin{aligned}
\sigma^{ES}_{ij} &= -V_{kj}P_{k,i} + \frac{1}{2}\left(E_k P_k + V_{kl}P_{k,l}\right)\delta_{ij} - D_j\phi_{,i} + \left(-\frac{1}{2}\varepsilon_0\phi_{,k}\phi_{,k} + \phi_{,k}P_k\right)\delta_{ij} \\
&= \sigma^M_{ij} + \tau^M_{ijm,m},
\end{aligned} \tag{3.91}$$

σ_{ij}^{M} and $\tau_{ijm,m}^{M}$ referring to the Maxwell stress and the electrostatic stress, respectively:

$$\sigma_{ij}^{M} = \frac{1}{2} E_k P_k \delta_{ij} - D_j \phi_{,j} + \phi_{,i} + \left(-\frac{1}{2} \varepsilon_0 \phi_{,k} \phi_{,k} + \phi_{,k} P_k \right) \delta_{ij} \tag{3.92a}$$

$$\tau_{ijm,m}^{M} = -V_{kj} P_{k,j} + \frac{1}{2} V_{kl} P_{k,l}. \tag{3.92b}$$

Let us assume that the mechanical displacement and polarization do not exist in the surrounding air environment. Then, we obtain the final boundary value problem by substituting Eqs. (3.90) and (3.87) into Eq. (3.86) and utilizing the arbitrariness of δu_i, $\delta \phi$ and δP_i resulting in

$$\left(\sigma_{ij} - \tau_{ijm,m} + \sigma_{ij}^{ES} \right)_{,j} + f_i = 0, \quad \text{in v} \tag{3.93a}$$

$$\sigma_{ij,j}^{ES} = 0, \quad \text{in v'} \tag{3.93b}$$

$$V_{ij,j} + E_i^{L} - \phi_{,i} + E_i^{ext} = 0, \quad \text{in v} \tag{3.93c}$$

$$-\varepsilon_0 \phi_{,ii} + P_{i,i} = \rho^{f}, \quad \text{in v} \tag{3.93d}$$

$$\phi_{,ii} = 0, \quad \text{in v'} \tag{3.93e}$$

$$u_i = \bar{u}_i, \quad \text{on } a^{u} \tag{3.94a}$$

$$\phi = \bar{\phi}, \quad \text{on } a^{\phi} \tag{3.94b}$$

$$\tilde{D}^{n}(u_i) = u_{i,l} n_l = \bar{u}_{i,l} n_l, \quad \text{on } a^{u,n} \tag{3.94c}$$

$$P_i = \bar{P}_i, \quad \text{on } a^{P} \tag{3.94d}$$

$$\left(\sigma_{ij} - \tau_{ijm,m} + [\![\sigma_{ij}^{ES}]\!] \right) n_j + \tilde{D}_l^{t}(n_l) n_m n_j \tau_{ijm} - \tilde{D}_j^{t}\left(n_m \tau_{ijm} \right) = t_i, \quad \text{on } a^{\sigma} \tag{3.95a}$$

$$\tau_{ijm} n_m n_j = 0, \quad \text{on } a^{\tau} \tag{3.95b}$$

$$n_i \left(-\varepsilon_0 [\![\phi_{,i}]\!] + P_i \right) = -\sigma^{*}, \quad \text{on } a^{\sigma^{*}} \tag{3.95c}$$

$$V_{ij} n_j = \bar{v}_i, \quad \text{on } a^{V} \tag{3.95d}$$

3.4.2.3 Formulation Based on the Gibbs Free Energy

An alternative formulation is based on the electric Gibbs free energy, that is equivalent to the electric enthalpy for an isothermal system. The electric Gibbs free energy is expressed in terms of the linear strain tensor ε_{ij}, the strain gradient $\varepsilon_{ij,k}$, electric field E_i and electric field gradient $E_{i,j}$ [50–52]:

$$
\begin{aligned}
G(\varepsilon_{ij}, \varepsilon_{ij,k}, E_i, E_{i,j}) = &-\frac{1}{2}\kappa_{ij}E_iE_j - \frac{1}{2}b_{ijkl}E_{i,j}E_{k,l} + \frac{1}{2}c_{ijkl}\varepsilon_{ij}\varepsilon_{kl} \\
&- e_{ijk}E_i\varepsilon_{jk} - \frac{1}{2}\mu_{ijkl}(E_k\varepsilon_{ij,l} - \varepsilon_{ij}E_{k,l})
\end{aligned}
\tag{3.96}
$$

Taking derivatives of the Gibbs free energy w.r.t. the four independent variables leads to the constitutive equations:

$$\sigma_{ij} = \frac{\partial G}{\partial \varepsilon_{ij}} = c_{ijkl}\varepsilon_{kl} + \frac{1}{2}\mu_{ijkl}E_{k,l} \tag{3.97a}$$

$$\tau_{ijk} = \frac{\partial G}{\partial \varepsilon_{ij,k}} = -\frac{1}{2}\mu_{ijkl}E_l \tag{3.97b}$$

$$D_i = -\frac{\partial G}{\partial E_i} = \kappa_{ij}E_j + \frac{1}{2}\mu_{ijkl}\varepsilon_{k,l} \tag{3.97c}$$

$$Q_{ij} = \frac{\partial G}{\partial E_{i,j}} = b_{ijkl}E_{k,l} - \frac{1}{2}\mu_{ijkl}\varepsilon_{kl}, \tag{3.97d}$$

where σ_{ij}, D_i, τ_{ijk} and V_{ij} indicate the (local) stress, electric displacement, higher-order stress and higher-order electric displacement, respectively while κ_{ij}, b_{ijkl}, c_{ijkl}, e_{ijk}, μ_{ijkl} are material tensors. In order to derive the final boundary values problem, we first take the first variation of the electric Gibbs free energy resulting in

$$\delta G = \sigma_{ij}\delta\varepsilon_{ij} + \tau_{ijk}\delta\varepsilon_{ij,k} - D_i\delta E_i + Q_{ij}\delta E_{i,j}, \tag{3.98}$$

Following the previous procedure, we define equilibrium from a variational principle

$$\delta \int_{v^*} G\,\mathrm{d}v^* = \delta W, \tag{3.99}$$

with the external work W. Instead of the contributions of the external electric field and higher-order electric field, higher-order surface charges $\bar{q}$ are prescribed on the boundary, so that the variation of δW reads

$$\delta W = \int_v \left(f_i\delta u_i + E_i^{\mathrm{ext}}\delta P_i\right)\mathrm{d}v + \int_{a^\sigma} t_i\delta u_i\,\mathrm{d}a + \int_{a^\tau} \bar{\tau}_i\tilde{D}^n(\delta u_i)\,\mathrm{d}a \tag{3.100}$$

$$+ \int_{a^Q} \bar{q}\tilde{D}^n(\delta\phi)\,\mathrm{d}a - \int_{a^\phi} \sigma^*\delta\phi\,\mathrm{d}a - \int_v \rho^f\delta\phi\,\mathrm{d}v\,. \tag{3.101}$$

In analogy to Eq.(3.89), the migratory variation of the electric field and electric field gradient are obtained as

$$\delta E_i = \delta_\phi E_i + \delta_u E_i = -\delta_\phi \phi_{,i} + E_{i,p}\delta u_p = -\delta_\phi \phi_{,i} + E_{p,i}\delta u_p, \tag{3.102a}$$

$$\delta E_{i,j} = \delta_\phi E_{i,j} + \delta_u E_{i,j} = \delta_\phi E_{i,j} + E_{i,jp}\delta u_p = \delta_\phi E_{i,j} + E_{i,pj}\delta u_p \tag{3.102b}$$

Substituting Eq. (3.102) into Eq. (3.99) and using Eq. (3.98) together with Gauss divergence theorem leads to

$$\begin{aligned}
\delta \int_v G\,\mathrm{d}v = &\int_v \left(-\sigma_{ij,j} + \tau_{ijk,kj} - \sigma^M_{ij,j} - \tau^M_{ijm,mj}\right)\delta u_i\,\mathrm{d}v \\
&+ \int_a \left(\sigma_{ij}n_j - \tau_{ijk,k}n_j + \sigma^M_{ij}n_j + \tau^M_{ijm,m}n_j\right. \\
&\left. + \tilde{D}^t_l(n_l)n_m n_j \tau_{ijm} - \tilde{D}^t_j\left(n_m\tau_{ijm}\right)\right)\delta u_i\,\mathrm{d}a \\
&+ \int_a \tau_{ijk}n_k n_j \tilde{D}^n(\delta u_i)\,\mathrm{d}a + \int_v \left(-D_{i,i} + D_{ij,ji}\right)\delta\phi\,\mathrm{d}v \\
&+ \int_a \left(D_i n_i - D_{ij,j}n_i + \tilde{D}^t_l(n_l)n_i n_j Q_{ij} - \tilde{D}^t_i(Q_{ij}n_j)\right)\delta\phi\,\mathrm{d}a \\
&+ \int_a Q_{ij}n_i n_j \tilde{D}^n(\delta\phi)\,\mathrm{d}a\,,
\end{aligned} \tag{3.103}$$

with the Maxwell stress σ^M_{ij} and higher-order electrostatic stress $\tau^M_{ijm,m}$, which are a direct outcome from the variational formulation:

$$\sigma^M_{ij} = E_i D_j - \frac{1}{2}E_k D_k \delta_{ij} \tag{3.104a}$$

$$\tau^M_{ijm,m} = E_{i,k}Q_{kj} - \frac{1}{2}E_{k,l}Q_{kl}\delta_{ij}. \tag{3.104b}$$

Note that Eq. (3.104a) and Eq. (3.92a) are identical. Substituting Eq. (3.103) and Eq. (3.100) into Eq. (3.99) leads to the following boundary value problem

$$\left(\sigma_{ij} - \tau_{ijm,m} + \sigma^{ES}_{ij}\right)_{,j} + f_i = 0, \quad \text{in v} \tag{3.105a}$$

$$\sigma^{ES}_{ij,j} = 0, \quad \text{in v'} \tag{3.105b}$$

$$\left(D_i - Q_{ij,j}\right)_{,i} = \rho^f, \quad \text{in v} \tag{3.105c}$$

$$\left(D_i - Q_{ij,j}\right)_{,i} = \phi_{,ii} = 0, \quad \text{in v'} \tag{3.105d}$$

$$u_i = \bar{u}_i, \quad \text{on } a^u \tag{3.106a}$$

$$\phi = \bar{\phi}, \quad \text{on } a^\phi \tag{3.106b}$$

$$\tilde{D}^n(u_i) = u_{i,l} n_l = \bar{u}_{i,l} n_l, \quad \text{on } a^{u,n} \tag{3.106c}$$

$$\tilde{D}^n(\phi) = \phi_{,l} n_l = \bar{\phi}_{,l} n_l, \quad \text{on } a^{\phi,n} \tag{3.106d}$$

$$\left(\sigma_{ij} - \tau_{ijm,m} + [\![\sigma_{ij}^{ES}]\!]\right) n_j + \tilde{D}_l^t(n_l) n_m n_j \tau_{ijm} - \tilde{D}_j^t\left(n_m \tau_{ijm}\right) = t_i, \quad \text{on } a^\sigma \tag{3.107a}$$

$$\tau_{ijk} n_k n_j = t\bar{a}u_i, \quad \text{on } a^\tau \tag{3.107b}$$

$$[\![D_i - Q_{ij,j}]\!] n_i + \tilde{D}_l^t(n_l) n_i n_j Q_{ij} - \tilde{D}_i^t(Q_{ij} n_j) = -\sigma^*, \quad \text{on } a^{\sigma^*} \tag{3.107c}$$

$$Q_{ij} n_i n_j = \bar{q}, \quad \text{on } a^Q \tag{3.107d}$$

Note that in vacuum the displacement and polarization do not exist degenerating $D_i - Q_{ij,j}$ to $-\phi_{,i}$, which also causes the jump $[\![D_i - Q_{ij,j}]\!]$ on the boundary a^ϕ.

3.4.2.4 Phase-Field Models for Fracture

Phase field models have been used extensively in the physics community for decades. They are particularly useful for solving moving boundary problems as demonstrated for instance in [53–56]. Instead of assuming a discrete boundary (or interface), which requires (a) a method to represent the boundary explicitly or implicitly (for instance using level sets) and (b) methods to track the moving boundary, phase field models smear the discrete boundary (or interface) over a certain region and solve an additional differential equation for the phase field to determine the (new) position of the (moving) boundary, which drastically simplifies the implementation. Phase field models have been used to solve phase transformation problems [57–59], electrochemical reactions [60–62], fracture mechanics problems [63–65] and domain evolution in ferroelectric and ferromagnetic materials [66–68]. A phase field model has been developed by Aland [69] to investigate biomembranes and inhomogeneities due to inserted proteins, absorbed molecules, or different compositions of lipids and cholesterol. Lázaro et al. [70] presented a phase-field formulation to simulate complex phenomena related to dynamics and morphology of biological membranes such as pearling, tubulation and RBCs hydrodynamics in microchannels. Rosolen et al. [71] presented an adaptive meshfree method based on the maximum-entropy approximants for simulating phase-field models of biomembranes based on Helfrich curvature elastic energy. Wang et al. [72] combined exponential time differencing Runge–Kutta approximations for time integration with spectral discretizations for spatial operators on regular meshes to solve the unconstrained and constrained Willmore flow problem using the phase field method. They also proposed a modified augmented Lagrange multiplier approach to avoid numerical instabilities caused by large penalty terms imposed on the system to penalize the violations in surface area and volume constraints. This subsection focuses on phase-field models for fracture, which have become very popular in recent years as it provides a thermodynamically consistent framework for material failure, which can be easily extended to a variety of problems including multi-field problems. We will present the approach presented by Miehe et al. [13, 73] though such approaches can be traced back to the 1990 where Francfort and Marigo [74] and Bourdin et al. [75] suggested a variational approach to fracture.

As mentioned before, the phase field model to fracture employs a monotonically increasing continuous scalar-valued phase field parameter $\phi(\mathbf{x}) \in [0, 1]$; $\phi = 0$ referring to the intact state while $\phi = 1$ indicates total fracture. Similar to gradient models, another differential equation is solved for the evolution of the phase field. The coupling to the mechanical field is realized by adding a stress degradation function to the energy in the bulk, which will finally yield a 'modified' equilibrium equation in strong form. In general, the stress-degradation function must satisfy the following properties [76]:

- $g(0) = 1$, since $\phi = 0$ represents a no-cracked zone.
- $g(1) = 0$, since $\phi = 1$ represents a completely damaged zone.
- $g(\phi) > 0$ if $\phi \neq 1$, prevents a fully damaged state.
- $g'(0) \neq 0$, since crack nucleation starts from a uniform initial state.
- $g'(1) = 0$, since the energy must converge to a finite value for a fully damaged state, i.e. $\phi = 1$.

Several stress degradation functions have been proposed in the literature including the 'classical' quadratic $g = (1-\phi)^2$ and cubic one $g = s\left[(1-\phi)^3 - (1-\phi)^2\right] + 3\,(1-\phi)^2 - 2\,(1-\phi)^3$, $s > 0$ determines the slope of g in the undamaged state which is commonly chosen close to 0. The quadratic degradation function finally leads to a formulation of the '$1-\phi$-type' as in classical scalar damage mechanics. However, it was shown for instance in [77] that a cubic stress degradation function yields more accurate results. This was demonstrated by solving only the phase field equation and testing how well the two different formulations can approximate a discrete crack surface.

In the variational theory to fracture, the total energy potential is additively decomposed into the bulk energy Ψ_e and the surface energy Ψ_c associated with the energy required for the formation of a crack, i.e.

$$\Psi = \Psi_e + \Psi_c \tag{3.108}$$

For an elastic isotropic solid, the bulk energy Ψ_e is expressed in terms of the strain energy density $\Psi_0(\boldsymbol{\epsilon})$, i.e.

$$\Psi_0(\boldsymbol{\epsilon}) = \frac{\lambda}{2}\mathrm{tr}^2(\boldsymbol{\epsilon}) + \mu\mathrm{tr}(\boldsymbol{\epsilon}^2) \tag{3.109}$$

where tr (•) denotes the trace of the tensor while λ and μ are the Lamé constants defined as:

$$\lambda = \frac{\nu E}{(1+\nu)(1-2\nu)}, \mu = \frac{E}{2(1+\nu)}, \tag{3.110}$$

E being the Young's modulus and ν the Poisson's ratio. In order to ensure that fracture occurs only under tension and not compression, the strain energy density Ψ_0 is additively decomposed into a positive and negative part:

$$\Psi_0^{\pm}(\boldsymbol{\epsilon}) = \frac{\lambda}{2} \langle \mathrm{tr}(\boldsymbol{\epsilon}) \rangle_{\pm}^2 + \mu \mathrm{tr}(\boldsymbol{\epsilon}_{\pm}^2). \tag{3.111}$$

with

$$\boldsymbol{\epsilon}_{\pm} = \sum_{I=1}^{d} \langle \epsilon_I \rangle_{\pm} \boldsymbol{n}_I \otimes \boldsymbol{n}_I, \tag{3.112}$$

$\langle \epsilon_I \rangle_{\pm}$ referring to the principal strains, i.e. the eigenvalues of the linear strain tensor, while $\boldsymbol{n}_I$ are the associated eigenvectors; the bracket operators $\langle \bullet \rangle_{\pm} = \frac{1}{2}(\bullet \pm | \bullet |)$ define the positive and negative projections, respectively. Hence, the bulk energy Ψ_e reads

$$\Psi_e(\boldsymbol{\epsilon}) = g(\phi)\Psi_0^{+}(\boldsymbol{\epsilon}) + \Psi_0^{-}(\boldsymbol{\epsilon}) \tag{3.113}$$

where the stress degradation function couples the mechanical field to the phase field. The energy to create new crack surfaces is defined as

$$\Psi_c = \int_{\Gamma_c} G_c \, d\Gamma = G_c \, A \tag{3.114}$$

where G_c designates the critical energy release rate and Γ_c (or A) is the crack surface. The key idea of the phase field approach to fracture is to transform the surface integral to a domain integral. Therefore, let us consider a one-dimensional bar with a crack located at $x = 0$. Then, one can define a phase field variable [13]

$$\phi(x) = \exp\left(\frac{-|x|}{l_0}\right). \tag{3.115}$$

which smears the discrete crack over the entire domain Ω and automatically satisfies the homogeneous ordinary differential equation

$$\phi''(x) - \frac{1}{l_0^2}\phi(x) = 0 \text{ in } \Omega, \tag{3.116}$$

subjected to the Dirichlet-type boundary conditions:

$$\begin{aligned}
&\phi(0) = 1\,,\ \phi'(0) = 0\,,\\
&\lim_{x\to\infty} \phi(x) = \lim_{x\to-\infty} \phi(x) = 0,\\
&\text{and } \lim_{x\to\infty} \phi'(x) = \lim_{x\to-\infty} \phi'(x) = 0.
\end{aligned} \tag{3.117}$$

This is the well-known second-order phase field model, which has been commonly implemented in the context of the finite element method due to the inherent C^0 requirement. The parameter l_0 is a length scale which determines the 'width' of the crack. Equation (3.116) is the strong form associated with the variational problem:

$$\phi = \mathrm{Arg}\left\{\inf I(\phi)\right\}, \tag{3.118}$$

with

$$\begin{aligned}
&I(\phi) = \frac{1}{2}\int_{\Omega} \left(\phi^2 + l_0^2|\nabla\phi|^2\right) d\Omega, \text{ where } d\Omega = \Gamma dx\,,\\
&\phi(0) = 1 \text{ and } \lim_{x\to\infty} \phi(x) = \lim_{x\to-\infty} \phi(x) = 0.
\end{aligned} \tag{3.119}$$

As a consequence, we define

$$\Gamma_2(\phi) = \frac{1}{l_0} I(\phi) = \frac{1}{2l_0}\int_{\Omega} \left(\phi^2 + l_0^2|\nabla\phi|^2\right) d\Omega. \tag{3.120}$$

The coefficients of $\phi(x)$ and its derivatives in (3.116) are chosen such that they satisfy (3.121).

$$\int_{-\infty}^{\infty} G_c \Gamma_n(\phi) dx \approx \int_{\Gamma} G_c \, d\Gamma. \tag{3.121}$$

Hence, the energy to create new crack surfaces Ψ_c is approximated as

$$\Psi_c = \int_{\Gamma} G_c \, d\Gamma = \frac{G_c}{2l_0}\int_{\Omega} \left(\phi^2 + l_0^2|\nabla\phi|^2\right) d\Omega. \tag{3.122}$$

An alternative to the second-order model is the fourth-order phase field model [78]. For the one-dimensional bar with a crack at $x = 0$, the phase field variable can be expressed by

$$\phi(x) = \exp\left(\frac{-|2x|}{l_0}\right)\left(1 + \frac{|2x|}{l_0}\right). \tag{3.123}$$

which is the solution of the homogeneous ordinary differential equation [78]:

$$\phi^{(4)}(x) - \frac{8}{l_0^2}\phi''(x) + \frac{16}{l_0^4}\phi(x) = 0 \text{ in } \Omega, \tag{3.124}$$

subjected to the Dirichlet-type boundary conditions in (3.117). Following the same procedure as for the second-order model, we define $\Gamma_4(\phi)$ as:

$$\Gamma_4(\phi) = \frac{1}{2l_0}\int_\Omega \left(\phi^2 + \frac{l_0^2}{2}|\nabla\phi|^2 + \frac{l_0^4}{16}(\Delta\phi)^2\right) d\Omega = \frac{1}{l_0} I(\phi), \tag{3.125}$$

with

$$I(\phi) = \frac{1}{2}\int_\Omega \left(\phi^2 + \frac{l_0^2}{2}|\nabla\phi|^2 + \frac{l_0^4}{16}(\Delta\phi)^2\right) d\Omega. \tag{3.126}$$

The coefficients of $\phi(x)$ and its derivatives in (3.124) are chosen such that they satisfy (3.121). Hence, for $d \in \{2, 3\}$, Ψ_c for the fourth-order phase field model is approximated as:

$$\Psi_c = \int_\Gamma G_c \, d\Gamma = \frac{G_c}{2l_0}\int_\Omega \left(\phi^2 + \frac{l_0^2}{2}|\nabla\phi|^2 + \frac{l_0^4}{16}(\Delta\phi)^2\right) d\Omega. \tag{3.127}$$

One important parameter is the length scale parameter l_0 which determines the sharpness of the crack. It can be regarded as a pure numerical parameter which ensures the well-posedness of the underlying boundary value problem but under certain conditions, it can be related to 'physical' parameters of the material. In [79], an analytical solution for the critical tensile stress, $\boldsymbol{\sigma}_{cr}$ that a one-dimensional bar can sustain under tension, is derived as:

$$\boldsymbol{\sigma}_{cr} = \frac{9}{16}\sqrt{\frac{EG_c}{3l_0}}. \tag{3.128}$$

Shrinking l_0 to zero leads apparently to a singularity. However, when solving the equation w.r.t. l_0, one obtains

$$l_0 = \frac{27EG_c}{256\boldsymbol{\sigma}_{cr}^2}. \tag{3.129}$$

relating the intrinsic length scale to the tensile strength, critical energy release rate and Young's modulus. Though the extension of this formulation in higher dimensions and complex mixed-mode fracture analysis is difficult, yet (3.129) gives a rough estimate on how to choose l_0.

3.4.2.5 Cahn-Hilliard Equation

The Cahn-Hilliard (CH) equation [80] describes the phase separation in two component systems such as two-phase fluids and the phase transition in alloys, glasses and polymeric solutions. In material science, the CH equation is closely related to the microstructure evolution and spinodal decomposition of mixing liquids. The patterns of the phase separations govern the microstructure of the material and significantly affect its properties such as strength, hardness or conductivity, to name a few. Phase transitions arise from thermodynamic processes of infinite molecules. The Cahn-Hilliard equation is a phenomenological model describing this thermodynamic process. Instead of considering sharp interfaces between phases, the CH equation introduces a thin transition region between different materials avoiding the need for complex boundary tracking and mesh moving algorithms [81]. In this sense, it shows similarities to the phase field approach discussed earlier. The CH equation is based on an interfacial energy of Ginzburg-Landau type. Due to its simplicity, the CH equation has been used in numerous applications including multiphase fluid flow [82], bacterial films [83], thin films [84], image inpainting and topology optimization [85], micro-structures with elastic inhomogeneity [86], phase separation in multi-component alloys [87], multi-field problems with finite strain elasticity [88] and tumor growth [89], to name a few.

The CH equation is a fourth-order non-linear parabolic partial differential equation, which has been solved with various computational approaches such as FDM [90–93], semi-implicit Fourier spectral method [94–96], mixed finite element methods [97, 98], Discontinuous Galerkin method [99] and Isogeometric Analysis (IGA) [4, 100, 101], to name a few. Finite difference methods have difficulties for complex (interior) boundaries while FEM based on Lagrange polynomials exhibit problems due to the higher order continuity requirement. Therefore, it is standard to convert the fourth-order CH equation into a coupled system of two second-order partial differential equations (PDEs) [97, 98, 102] requiring only C^0 continuity. These coupled PDEs are commonly solved with mixed FE formulations at the cost of additional unknowns. Ensuring stability is another obstacle. Discontinuous Galerkin methods as suggested in [99] avoid the need for C^1 basis functions. However, first- and second-order partial derivatives need to be calculated on each inter-element boundary, which requires jump and average operators. Some meshfree and Isogeometric Analysis (IGA) formulations [100, 103, 104] are higher order continuous and therefore well suited for the solution of the CH equation. However, ensuring higher-order continuity at patch boundaries still remains a challenging in IGA, especially for complex geometries.

The CH equation with boundary conditions can be written as [99]

$$\partial_t c = \nabla \cdot \Big(D(c)\nabla\mu\Big),\ \mu = \frac{\delta E}{\delta c} = -\epsilon^2\nabla^2 c + F'(c) \text{ in } \Omega \tag{3.130a}$$

$$\nabla c \cdot n = D(c)\nabla\mu \cdot n = 0 \text{ on } \partial\Omega \tag{3.130b}$$

$$c(\boldsymbol{x}, 0) = c_0(\boldsymbol{x}) \text{ in } \Omega, \tag{3.130c}$$

$c \in (-1, 1)$ denoting the chemical concentration, $D(c)$ is the concentration dependent mobility, ϵ controls the width of the interface of two immiscible phases, μ is the chemical potential, $F'(c)$ refers to the derivative of the double-well potential with respect to the concentration and $c_0(\boldsymbol{x})$ indicates the initial distribution of the chemical concentration. Note that the concentration dependent mobility $D(c)$ is also sometimes assumed to be constant for sake of simplicity. The total free energy $E(c)$ is given by

$$E(c) = \int_\Omega (\epsilon^2|\nabla c|^2 + F(c)). \tag{3.131}$$

The concentration can be regarded as volume- or mass fraction in the mixture of two components. The volume fraction is equivalent to the mole fraction in materials science, where a mole refers to 6.02252×10^{23} molecules (Avogadro's number of molecules). Two common double-well potential functions are the Ginzburg-Landau double-well potential given by

$$F(c) = \frac{1}{4}\left(c^2 - 1\right)^2, \quad F'(c) = c^3 - c \tag{3.132}$$

and the logarithmic energy functional [105, 106], which is expressed as

$$F_{log}(c) = \frac{\Theta}{2}(1 - c^2) + \frac{1}{2}\Big((1 - c)\ln(1 - c) + (1 + c)\ln(1 + c)\Big) \tag{3.133}$$

$$F'_{log}(c) = -\Theta c + \frac{1}{2}\ln\frac{1 + c}{1 - c}, \tag{3.134}$$

Θ denoting the temperature, or more precisely the scaled temperature with $\Theta > 1$ being a requirement for the double-well potential. Figure 3.8 shows that the polynomial potential energy reaches the minimum at $c = \pm 1$ and the minimum of the logarithmic potential (i.e. $\Theta = 4/3$) is 0.6097 at $c = \pm 0.829$.

The parameter ϵ is related to the transient layer thickness. Let us consider the CH equation based on the polynomial potential $F(c) = \frac{1}{4}(c^2 - 1)^2$ in 1D. The steady distribution of the concentration requires the chemical potential $\mu(c) = \frac{\delta E}{\delta c} = c^3 - c - \epsilon^2 c'' = 0$. Under the boundary conditions $c_0(x \to \pm\infty) = \pm 1$ [107, 108], the exact solution can be written as

$$c(x) = \tanh\left(\frac{x}{\sqrt{2}\epsilon}\right) \tag{3.135}$$

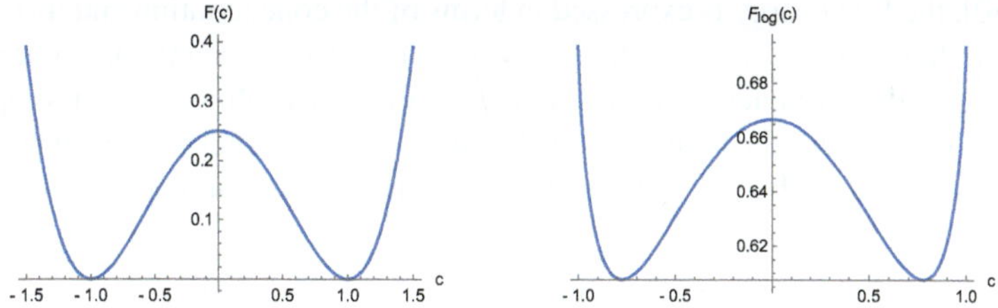

Fig. 3.8 (**a**) Polynomial potential, (**b**) Logarithmic potential $\Theta = 4/3$

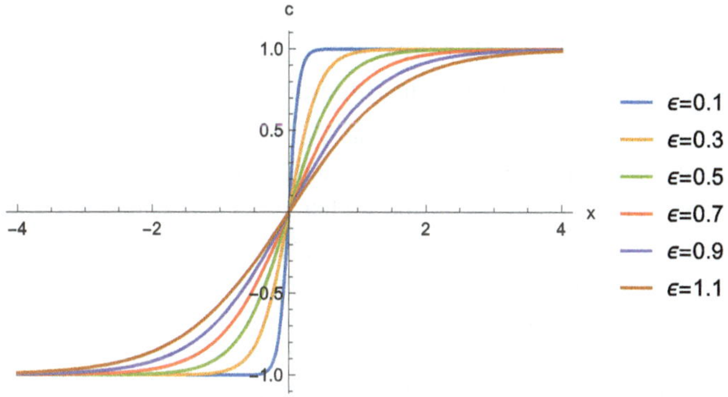

Fig. 3.9 The profiles of $\tanh\left(\frac{x}{\sqrt{2}\epsilon}\right)$

Figure 3.9 illustrates the profile of the exact solution with different values for ϵ. Increasing ϵ apparently increases the interface thickness. The total interface energy in 1D is given as

$$E(c) = \int_{-\infty}^{\infty} \frac{3}{4} \operatorname{sech}\left(\frac{x}{\sqrt{2}\epsilon}\right)^4 dx = \sqrt{2}\epsilon. \tag{3.136}$$

In order to reduce the total interface energy, ϵ should be small but it is restricted by the underlying discretization in practice. It can be shown that the weak form of Eq. (3.130a) in Ω can be written as

$$\begin{aligned} &\int_{\Omega} \Big(\partial_t c - \nabla \cdot (D(c)\nabla\mu)\Big)\delta c \, dV \\ &= \int_{\Omega} \Big(\partial_t c \, \delta c + \nabla\delta c \cdot (D(c)\nabla\mu)\Big) \, dV - \int_{\partial\Omega} D(c)\nabla\mu \cdot \boldsymbol{n}\delta c \, dS \\ &= \int_{\Omega} \partial_t c \, \delta c - \mu \, \nabla \cdot (D(c)\nabla\delta c) \, dV - \int_{\partial\Omega} D(c)\nabla\mu \cdot \boldsymbol{n}\delta c \, dS + \int_{\partial\Omega} \nabla\delta c \cdot \boldsymbol{n} D(c)\mu \, dS. \end{aligned} \tag{3.137}$$

In [80], the free energy is expressed in terms of the concentration and its (higher-order) gradients, i.e. $E(c) = f(c, \nabla c, \nabla^2 c, ...)$. Let us account for higher order terms $\nabla^2 c$ in the free energy, i.e. $E(c) = f(c, \nabla c, \nabla^2 c)$. Moreover, for simplicity, we only consider periodic boundary conditions without internal solid-wall boundary conditions so that the boundary terms vanish in the integration by parts yielding

$$E(c) = \int_\Omega \left(\frac{\epsilon^2}{2} (\nabla c \cdot \nabla c + \ell^2 \nabla^2 c : \nabla^2 c) + \frac{1}{4}(c^2 - 1)^2 \right) dV, \qquad (3.138)$$

where ℓ is the non-local characteristic length scale. Equation (3.138) is the simplest version of the higher order CH equation since it introduces only one additional parameter ℓ. For higher order terms, $\ell^2 \nabla^2 c : \nabla^2 c$ should be replaced by $g_{ijkl} \frac{\partial^2 c}{\partial x_i \partial x_j} \frac{\partial^2 c}{\partial x_k \partial x_l}$, g_{ijkl} denoting the 4th-order material tensor. The magnitude order of $\nabla^2 c$ is proportional to $1/h^2$ while $\nabla c \propto 1/h$, h being the characteristic length scale of the model. The parameter ℓ determines the influence of the higher order terms on $|\nabla c|$. When $\ell = 0$, the higher order CH equation reduces to the 'traditional' CH equation. When $\ell = \epsilon$, $\frac{\epsilon^2}{2} \ell^2 \nabla^2 c : \nabla^2 c$ becomes $\frac{\epsilon^2}{2} \nabla c \cdot \nabla c$. The parameter ℓ governs the non-locality, which might be significant at smaller length scales. The higher order CH equation is used to model physical phenomena such as the faceting of growing crystal surfaces [109], self-arranging nano-surfaces on crystalline substrates, quantum dots [110] or the separation of oil-water-surfactant mixtures [111–113]. The higher order CH equation can be derived from the nonlocal CH equation using certain kernel functions [114, 115]. The existence and uniqueness of the solution of the 6th order CH equation has been proven in [116–118]. Nonetheless, numerical studies employing the 6th-order CH equation are rare. Boyarkin et al. [119] studied the sixth order CH equation using a C^0 interior penalty discontinuous Galerkin approximation. In the scheme, the 6th order CH equation is converted into a second order and a fourth order PDE.

It can be shown that the chemical potential is given by

$$\begin{aligned} \mu :=& \delta_c E = c^3 - c - \epsilon^2 \Delta c + \epsilon^2 \ell^2 \nabla \cdot (\nabla \cdot (\nabla \nabla c)) \\ =& c^3 - c - \epsilon^2 \Delta c + \epsilon^2 \ell^2 \Delta^2 c \end{aligned} \qquad (3.139)$$

where we used $\nabla(\Delta c) = \nabla \cdot (\nabla^2 c)$ and $\nabla \cdot (\nabla \cdot (\nabla \nabla c)) = \Delta^2 c$. The strong form of the higher order CH equation with constant mobility reads

$$\frac{c}{t} = D \Delta (c^3 - c - \epsilon^2 \Delta c + \epsilon^2 \ell^2 \Delta^2 c). \qquad (3.140)$$

Finally, the weak form is given by

$$
\begin{aligned}
\mathscr{F}(c,\delta c) &= \int_\Omega (\partial_t c - D\nabla\cdot\nabla\mu)\delta c\, dV \\
&= \int_\Omega \Big(\partial_t c\,\delta c + \nabla\delta c\cdot(D\nabla\mu)\Big)\, dV \\
&= \int_\Omega \Big(\partial_t c\,\delta c - D\Delta\delta c\,\mu\Big)\, dV \\
&= \int_\Omega \Big(\partial_t c\,\delta c - D\Delta\delta c\,(c^3 - c - \epsilon^2\Delta c) - D\epsilon^2\ell^2\Delta\delta c\nabla\cdot(\nabla\Delta c))\Big)\, dV \\
&= \int_\Omega \Big(\partial_t c\,\delta c - D\Delta\delta c\,(c^3 - c - \epsilon^2\Delta c) + D\epsilon^2\ell^2\nabla(\Delta\delta c)\cdot(\nabla(\Delta c))\Big)\, dV.
\end{aligned}
\tag{3.141}
$$

which requires third-order partial derivatives.

References

1. París, F., Cañas, J.: Boundary Element Method: Fundamentals and Applications. Oxford Science Publications. Oxford University Press, Oxford (1997). ISBN: 0198565372, 9780198565376
2. Chorin, A.J.: Numerical solution of the Navier-Stokes equations. Math. Comput. **22**(104), 745–762 (1968). 10.1090/S0025-5718-1968-0242392-2
3. Chorin, A.J.: On the convergence of discrete approximations to the Navier Stokes equations. Math. Comput. **23**(106), 341–353 (1969). 10.1090/S0025-5718-1969-0242393-5
4. Hughes, T.J., Cottrell, J.A., Bazilevs, Y.: Isogeometric analysis: CAD, finite elements, NURBS, exact geometry and mesh refinement. Comput. Methods Appl. Mech. Eng. **194**(39–41), 4135–4195 (2005). 10.1016/j.cma.2004.10.008
5. Cottrell, J.A., Reali, A., Bazilevs, Y., Hughes, T.J.: Isogeometric analysis of structural vibrations. Comput. Methods Appl. Mech. Eng. **195**(41–43), 5257–5296 (2006). 10.1016/j.cma.2005.09.027
6. Cottrell, J.A., Hughes, T.J., Reali, A.: Studies of refinement and continuity in isogeometric structural analysis. Comput. Methods Appl. Mech. Eng. **196**(41–44), 4160–4183 (2007). 10.1016/j.cma.2007.04.007
7. Liu, G.-R.: Meshfree Methods: Moving Beyond the Finite Element Method. Taylor & Francis (2009). 10.1201/9781420082104
8. Rabczuk, T., Xiao, S.P., Sauer, M.: Coupling of mesh-free methods with finite elements: basic concepts and test results. Commun. Numer. Methods Eng. **22**(10), 1031–1065 (2006). 10.1002/cnm.871
9. Gu, Y., Zhang, L.: Coupling of the meshfree and finite element methods for determination of the crack tip fields. Eng. Fract. Mech. **75**(5), 986–1004 (2008). 10.1016/j.engfracmech.2007.05.003
10. Li, W., Nguyen-Thanh, N., Zhou, K.: Geometrically nonlinear analysis of thin-shell structures based on an isogeometric-meshfree coupling approach. Comput. Methods Appl. Mech. Eng. **336**, 111–134 (2018). 10.1016/j.cma.2018.02.018

11. Nguyen-Thanh, N., Huang, J., Zhou, K.: An isogeometric-meshfree coupling approach for analysis of cracks. Int. J. Numer. Methods Eng. **113**(10), 1630–1651 (2018). 10.1002/nme.5713
12. Griffith, A.A.: VI. The phenomena of rupture and flow in solids. Philos. Trans. R. Soc. Lond. Ser. A, containing papers of a mathematical or physical character **221**(582–593), 163–198 (1921). 10.1098/rsta.1921.0006
13. Miehe, C., Welschinger, F., Hofacker, M.: Thermodynamically consistent phase-field models of fracture: variational principles and multi-field FE implementations. Int. J. Numer. Methods Eng. **83**(10), 1273–1311 (2010). 10.1002/nme.2861
14. Ball, J.M., Chen, G.Q.G.: Entropy and convexity for nonlinear partial differential equations. Philos. Trans. R. Soc. A Math. Phys. Eng. Sci. **371**(2005), 20120340 (2013). 10.1098/rsta.2012.0340
15. Zagoskin, A., Rakhmanov, A., Savel'Ev, S., Nori, F.: Quantum metamaterials: Electromagnetic waves in Josephson qubit lines. Phys. Status Sol. (b) **246**(5), 955–960 (2009). 10.1002/pssb.200881568
16. Okamoto, H.: Models and special solutions of the Navier–Stokes equations. In: Handbook of Mathematical Analysis in Mechanics of Viscous Fluids. Springer, New York (2018). 10.1007/978-3-319-13344-7_14
17. Spencer, A.J.M.: Continuum Mechanics. Courier Corporation, Mineola (2004). https://books.google.de/books?id=w9OV8P2ELVwC
18. Holzapfel, G.A.: Nonlinear Solid Mechanics: A Continuum Approach for Engineering Science. Kluwer Academic Publishers, Dordrecht (2002). ISBN: 9780471823049. https://www.wiley.com/en-us/Nonlinear%2BSolid%2BMechanics%3A%2BA%2BContinuum%2BApproach%2Bfor%2BEngineering-p-9780471823193
19. Ball, J.M.: Convexity conditions and existence theorems in nonlinear elasticity. Arch. Rational Mech. Anal. **63**(4), 337–403 (1976). 10.1007/BF00279992
20. Klein, D.K., Fernández, M., Martin, R.J., Neff, P., Weeger, O.: Polyconvex anisotropic hyperelasticity with neural networks. J. Mech. Phys. Solids **159**, 104703 (2022). 10.1016/j.jmps.2021.104703
21. Korelc, J., Wriggers, P.: Automation of Finite Element Methods. Springer, Cham (2016). 10.1007/978-3-319-39005-5
22. Korelc, J.: Automation of primal and sensitivity analysis of transient coupled problems. Comput. Mech. **44**(5), 631–649 (2009). 10.1007/s00466-009-0395-2
23. Alnæs, M.S., Logg, A., Ølgaard, K.B., Rognes, M.E., Wells, G.N.: Unified form language: a domain-specific language for weak formulations of partial differential equations. ACM Trans. Math. Softw. (TOMS) **40**(2), 1–37 (2014). 10.1145/2566630
24. Hughes, T.J., Winget, J.: Finite rotation effects in numerical integration of rate constitutive equations arising in large-deformation analysis. Int. J. Numer. Methods Eng. **15**(12), 1862–1867 (1980). 10.1002/nme.1620151210
25. Simo, J., Miehe, C.: Associative coupled thermoplasticity at finite strains: formulation, numerical analysis and implementation. Comput. Methods Appl. Mech. Eng. **98**(1), 41–104 (1992). 10.1016/0045-7825(92)90170-O
26. Simo, J., Ortiz, M.: A unified approach to finite deformation elastoplastic analysis based on the use of hyperelastic constitutive equations. Comput. Methods Appl. Mech. Eng. **49**(2), 221–245 (1985). 10.1016/0045-7825(85)90061-1
27. Simo, J.C.: Algorithms for static and dynamic multiplicative plasticity that preserve the classical return mapping schemes of the infinitesimal theory. Comput. Methods Appl. Mech. Eng. **99**(1), 61–112 (1992). 10.1016/0045-7825(92)90123-2
28. Auricchio, F., Taylor, R.L.: Two material models for cyclic plasticity: nonlinear kinematic hardening and generalized plasticity. Int. J. Plast. **11**(1), 65–98 (1995). 10.1016/0749-6419(94)00039-5
29. Gasser, T.C., Holzapfel, G.A.: Modeling the propagation of arterial dissection. Eur. J. Mech. A/Sol. **25**(4), 617–633 (2006). 10.1016/j.euromechsol.2006.05.004

30. Bischoff, J.E., Arruda, E.M., Grosh, K.: Finite element simulations of orthotropic hyperelasticity. Finite Elem. Anal. Des. **38**(10), 983–998 (2002). 10.1016/S0168-874X(02)00089-6
31. Mashkevich, V., Tolpygo, K.: Electrical, optical and elastic properties of diamond type crystals. Sov. Phys. JETP **5**(3), 435–439 (1957). http://jetp.ras.ru/cgi-bin/dn/e_005_03_0435.pdf
32. Deng, Q., Kammoun, M., Erturk, A., Sharma, P.: Nanoscale flexoelectric energy harvesting. Int. J. Solids Struct. **51**(18), 3218–3225 (2014). 10.1016/j.ijsolstr.2014.05.018
33. Bhaskar, U.K., Banerjee, N., Abdollahi, A., Wang, Z., Schlom, D.G., Rijnders, G., et al.: A flxoelectric microelectromechanical system on silicon. Nat. Nanotechnol. **11**(3), 263–266 (2016). 10.1038/nnano.2015.260
34. Abdollahi, A., Arias, I.: Constructive and destructive interplay between piezoelectricity and flexoelectricity in flexural sensors and actuators. J. Appl. Mech. **82**(12) (2015). 10.1115/1.4031333
35. Mbarki, R., Haskins, J., Kinaci, A., Cagin, T.: Temperature dependence of flexoelectricity in BaTiO3 and SrTiO3 perovskite nanostructures. Phys. Lett. A **378**(30–31), 2181–2183 (2014). 10.1016/j.physleta.2014.05.011
36. Kwon, S.R., Huang, W., Shu, L., Yuan, F.-G., Maria, J.-P., Jiang, X.: Flexoelectricity in barium strontium titanate thin film. Appl. Phys. Lett. **105**(14) (2014). 10.1063/1.4898139
37. Narvaez, J.: Flexoelectricity in Single Crystals. Universitat Autònoma de Barcelona, Barcelona (2016)
38. Yudin, P., Ahluwalia, R., Tagantsev, A.: Upper bounds for flexoelectric coefficients in ferroelectrics. Appl. Phys. Lett. **104**(8), 082913 (2014). 10.1063/1.4865208
39. Yudin, P., Tagantsev, A.: Fundamentals of flexoelectricity in solids. Nanotechnology **24**(43), 432001 (2013). 10.1088/0957-4484/24/43/432001
40. Kvasov, A., Tagantsev, A.K.: Dynamic flexoelectric effect in perovskites from first-principles calculations. Phys. Rev. B **92**(5), 054104 (2015). 10.1103/PhysRevB.92.054104
41. Resta, R.: Towards a bulk theory of flexoelectricity. Phys. Rev. Lett. **105**(12), 127601 (2010). 10.1103/PhysRevLett.105.127601
42. Kuang, Z.B.: Some problems in electrostrictive and magnetostrictive materials. Acta Mech. Sol. Sin. **20**(3), 219–227 (2007). 10.1007/s10338-007-0726-9
43. Kuang, Z.-B.: Some variational principles in elastic dielectric and elastic magnetic materials. Eur. J. Mech. A/Solids **27**(3), 504–514 (2008). 10.1016/j.euromechsol.2007.10.001
44. Kuang, Z.-B.: Internal energy variational principles and governing equations in electroelastic analysis. Int. J. Solids Struct. **46**(3–4), 902–911 (2009). 10.1016/j.ijsolstr.2008.10.001
45. Kuang, Z.-B.: Theory of Electroelasticity. Springer, Berlin (2014). 10.1007/978-3-642-36291-0
46. Toupin, R.A.: The elastic dielectric. J. Rational Mech. Anal. **5**(6), 849–915 (1956). 10.1512/iumj.1956.5.55033
47. Hu, S., Shen, S.: Variational principles and governing equations in nanodielectrics with the flexoelectric effect. Sci. China Phys. Mech. Astron. **53**(8), 1497–1504 (2010). 10.1007/s11433-010-4039-5
48. Mindlin, R.D.: Polarization gradient in elastic dielectrics. Int. J. Solids Struct. **4**(6), 637–642 (1968). 10.1115/1.3423753
49. Mindlin, R.D., Eshel, N.: On first strain-gradient theories in linear elasticity. Int. J. Solids Struct. **4**(1), 109–124 (1968). 10.1016/0020-7683(68)90036-X
50. Abdollahi, A., Peco, C., Millan, D., Arroyo, M., Arias, I.: Computational evaluation of the flexoelectric effect in dielectric solids. J. Appl. Phys. **116**(9), 093502 (2014). 10.1063/1.4893974
51. Abdollahi, A., Millán, D., Peco, C., Arroyo, M., Arias, I.: Revisiting pyramid compression to quantify flexoelectricity: a three-dimensional simulation study. Phys. Rev. B **91**(10), 104103 (2015). 10.1103/PhysRevB.91.104103
52. Deng, F., Deng, Q., Yu, W., Shen, S.: Mixed finite elements for flexoelectric solids. J. Appl. Mech. **84**(8) (2017). 10.1115/1.4036939

53. Langer, J.: Models of pattern formation in first-order phase transitions. Directions in Condensed Matter Physics: Memorial Volume in Honor of Shang-Keng Ma. World Scientific, Singapore, pp. 165–186 (1986). 10.1142/9789814415309_0005
54. Moelans, N., Blanpain, B., Wollants, P.: An introduction to phase-field modeling of microstructure evolution. Calphad **32**(2), 268–294 (2008). 10.1016/j.calphad.2007.11.003
55. Anderson, D.M., McFadden, G.B., Wheeler, A.A.: Diffuse-interface methods in fluid mechanics. Annu. Rev. Fluid Mech. **30**(1), 139–165 (1998). 10.1146/annurev.fluid.30.1.139
56. Chen, L.-Q., Khachaturyan, A.: Computer simulation of structural transformations during precipitation of an ordered intermetallic phase. Acta Metall. Mater. **39**(11), 2533–2551 (1991). 10.1016/0956-7151(91)90069-D
57. Mamivand, M., Zaeem, M.A., El Kadiri, H.:. A review on phase field modeling of martensitic phase transformation. Comput. Mater. Sci. **77**, 304–311 (2013). 10.1016/j.commatsci.2013.04.059
58. Zhu, B., Chen, H., Militzer, M.: Phase-field modeling of cyclic phase transformations in low-carbon steels. Comput. Mater. Sci. **108**, 333–341 (2015). 10.1016/j.commatsci.2015.01.023
59. Yeon, D.-H., Cha, P.-R., Kim, J.-H., Grant, M., Yoon, J.-K.: A phase field model for phase transformation in an elastically stressed binary alloy. Model. Simul. Mater. Sci. Eng. **13**(3), 299 (2005). 10.1088/0965-0393/13/3/001
60. Zhao, Y., Xu, B.-X., Stein, P., Gross, D.: Phase-field study of electrochemical reactions at exterior and interior interfaces in Li-ion battery electrode particles. Comput. Methods Appl. Mech. Eng. **312**, 428–446 (2016). 10.1016/j.cma.2016.04.033
61. Lin, C., Ruan, H., Shi, S.-Q.: Phase field study of mechanico-electrochemical corrosion. Electrochim. Acta **310**, 240–255 (2019). 10.1016/j.electacta.2019.04.076
62. Mu, Z., Guo, Z., Lin, Y.-H.: Simulation of 3-D lithium dendritic evolution under multiple electrochemical states: a parallel phase field approach. Energy Storage Mater. (2020). 10.1016/j.ensm.2020.04.011
63. Miehe, C., Aldakheel, F., Raina, A.: Phase field modeling of ductile fracture at finite strains: a variational gradient-extended plasticity-damage theory. Int. J. Plast. **84**, 1–32 (2016). 10.1016/j.ijplas.2016.04.011
64. Zhang, X., Krischok, A., Linder, C.: A variational framework to model diffusion induced large plastic deformation and phase field fracture during initial two-phase lithiation of silicon electrodes. Comput. Methods Appl. Mech. Eng. **312**, 51–77 (2016). 10.1016/j.cma.2016.05.007
65. Miehe, C., Dal, H., Schänzel, L.-M., Raina, A.: A phase-field model for chemomechanical induced fracture in lithium-ion battery electrode particles. Int. J. Numer. Methods Eng. **106**(9), 683–711 (2016). 10.1002/nme.5133
66. Choudhury, S., Li, Y., Krill Iii, C., Chen, L.-Q.: Phase-field simulation of polarization switching and domain evolution in ferroelectric polycrystals. Acta Mater. **53**(20), 5313–5321 (2005). 10.1016/j.actamat.2005.07.040
67. Schrade, D., Mueller, R., Xu, B., Gross, D.: Domain evolution in ferroelectric materials: a continuum phase field model and finite element implementation. Comput. Methods Appl. Mech. Eng. **196**(41–44), 4365–4374 (2007). 10.1016/j.cma.2007.05.010
68. Yi, M., Xu, B.-X.: A constraint-free phase field model for ferromagnetic domain evolution. Proc. R. Soc. A Math. Phys. Eng. Sci. **470**(2171), 20140517 (2014). 10.1098/rspa.2014.0517
69. Aland, S.: Phase field models for two-phase flow with surfactants and biomembranes. In: Transport Processes at Fluidic Interfaces, pp. 271–290. Springer, Cham (2017). 10.1007/978-3-319-56602-3_11
70. Lázaro, G.R., Pagonabarraga, I., Hernández-Machado, A.: Phase-field theories for mathematical modeling of biological membranes. Chem. Phys. Lipids **185**, 46–60 (2015). Membrane mechanochemistry: From the molecular to the cellular scale. ISSN: 0009-3084. 10.1016/j.chemphyslip.2014.08.001
71. Rosolen, A., Peco, C., Arroyo, M.: An adaptive meshfree method for phasefield models of biomembranes. Part I: Approximation with maximum-entropy basis functions. J. Comput. Phys. **249**, 303–319 (2013). 10.1016/j.jcp.2013.04.046

72. Wang, X., Ju, L., Du, Q.: Efficient and stable exponential time differencing Runge–Kutta methods for phase field elastic bending energy models. J. Comput. Phys. **316**, 21–38 (2016). 10.1016/j.jcp.2016.04.004
73. Miehe, C., Hofacker, M., Welschinger, F.: A phase field model for rate-independent crack propagation:Robust algorithmic implementation based on operator splits. Comput. Methods Appl. Mech. Eng. **199**(45–48), 2765–2778 (2010). 10.1016/j.cma.2010.04.011
74. Francfort, G., Marigo, J.-J.: Revisiting brittle fracture as an energy minimization problem. J. Mech. Phys. Solids **46**(8), 1319–1342 (1998). 10.1016/S0022-5096(98)00034-9
75. Bourdin, B., Francfort, G.A., Marigo, J.-J.: Numerical experiments in revisited brittle fracture. J. Mech. Phys. Solids **48**(4), 797–826 (2000). 10.1016/S0022-5096(99)00028-9
76. Sargado, J.M., Keilegavlen, E., Berre, I., Nordbotten, J.M.: High-accuracy phase-field models for brittle fracture based on a new family of degradation functions. J. Mech. Phys. Solids **111**, 458–489 (2018). 10.1016/j.jmps.2017.10.015
77. Goswami, S., Anitescu, C., Rabczuk, T.: Adaptive fourth-order phase field analysis for brittle fracture. Comput. Methods Appl. Mech. Eng. **361**, 112808 (2020). 10.1016/j.cma.2019.112808
78. Borden, M.J., Hughes, T.J., Landis, C.M., Verhoosel, C.V.: A higher-order phase-field model for brittle fracture: formulation and analysis within the isogeometric analysis framework. Comput. Methods Appl. Mech. Eng. **273**, 100–118 (2014). 10.1016/j.cma.2014.01.016
79. Bažant, Z.P., Pijaudier–Cabot, G.: Measurement of characteristic length of nonlocal continuum. J. Eng. Mech. **115**(4), 755–767 (1989). 10.1061/(ASCE)0733-9399(1989)115:4(755)
80. Cahn, J.W., Hilliard, J.E.: Free energy of a nonuniform system. I. Interfacial free energy. J. Chem. Phys. **28**(2), 258–267 (1958). 10.1063/1.1744102
81. Valizadeh, N., Rabczuk, T.: Isogeometric analysis for phase-field models of geometric PDEs and high-order PDEs on stationary and evolving surfaces. Comput. Methods Appl. Mech. Eng. **351**, 599–642 (2019). 10.1016/j.cma.2019.03.043
82. Cahn, J.W., Hilliard, J.E.: Free energy of a nonuniform system. III. Nucleation in a two-component incompressible fluid. J. Chem. Phys. **31**(3), 688–699 (1959). 10.1063/1.1730447
83. Klapper, I., Dockery, J.: Role of cohesion in the material description of biofilms. Phys. Rev. E **74**(3), 031902 (2006). 10.1103/PhysRevE.74.031902
84. Oron, A., Davis, S.H., Bankoff, S.G.: Long-scale evolution of thin liquid films. Rev. Mod. Phys. **69**(3), 931 (1997). 10.1103/RevModPhys.69.931
85. Zhou, S., Wang, M.Y.: Multimaterial structural topology optimization with a generalized Cahn-Hilliard model of multiphase transition. Struct. Multidiscip. Optim. **33**(2), 89 (2007). 10.1007/s00158-006-0035-9
86. Kim, J., Lee, S., Choi, Y., Lee, S.-M., Jeong, D.: Basic principles and practical applications of the Cahn-Hilliard equation. Math. Prob. Eng. **2016** (2016). 10.1155/2016/9532608
87. Boyanova, P., Neytcheva, M.: Efficient numerical solution of discrete multicomponent Cahn-Hilliard systems. Comput. Math. Appl. **67**(1), 106–121 (2014). 10.1016/j.camwa.2013.10.013
88. Areias, P., Samaniego, E., Rabczuk, T.: A staggered approach for the coupling of Cahn–Hilliard type diffusion and finite strain elasticity. Comput. Mech. **57**(2), 339–351 (2016). 10.1007/s00466-015-1235-1
89. Mohammadi, V., Dehghan, M.: Simulation of the phase field Cahn-Hilliard and tumor growth models via a numerical scheme: Element-free Galerkin method. Comput. Methods Appl. Mech. Eng. **345**, 919–950 (2019). 10.1016/j.cma.2018.11.019
90. Sun, Z.Z.: A second-order accurate linearized difference scheme for the two-dimensional Cahn-Hilliard equation. Math. Comput. **64**(212), 1463–1471 (1995). 10.1090/S0025-5718-1995-1308465-4
91. Furihata, D.: A stable and conservative finite difference scheme for the Cahn-Hilliard equation. Numer. Math. **87**(4), 675–699 (2001). 10.1007/PL00005429
92. Lee, H.G., Kim, J.: A second-order accurate non-linear difference scheme for the N-component Cahn-Hilliard system. Phys. A Stat. Mech. Appl. **387**(19–20), 4787–4799 (2008). 10.1016/j.physa.2008.03.023

93. Diegel, A.E., Wang, C., Wise, S.M.: Stability and convergence of a second-order mixed finite element method for the Cahn-Hilliard equation. IMA J. Numer. Anal. **36**(4), 1867–1897 (2016). 10.1093/imanum/drv065
94. Zhu, J., Chen, L.-Q., Shen, J., Tikare, V.: Coarsening kinetics from a variable-mobility Cahn-Hilliard equation: application of a semi-implicit Fourier spectral method. Phys. Rev. E **60**(4), 3564 (1999). 10.1103/PhysRevE.60.3564
95. He, Z., Honeycutt, C.W., Xing, B., McDowell, R.W., Pellechia, P.J., Zhang, T.: Solid-state Fourier transform infrared and 31P nuclear magnetic resonance spectral features of phosphate compounds. Soil Sci. **172**(7), 501–515 (2007). 10.1097/SS.0b013e318053dba0
96. Shen, J., Tang, T., Wang, L.-L.: Spectral Methods: Algorithms, Analysis and Applications, vol. 41. Springer Science & Business Media, New York (2011). 10.1007/978-3-540-71041-7
97. Elliott, C.M., French, D.A., Milner, F.: A second order splitting method for the Cahn-Hilliard equation. Numer. Math. **54**(5), 575–590 (1989). 10.1007/BF01396363
98. Feng, X., Prohl, A.: Error analysis of a mixed finite element method for the Cahn-Hilliard equation. Numer. Math. **99**(1), 47–84 (2004). 10.1007/s00211-004-0546-5
99. Wells, G.N., Kuhl, E., Garikipati, K.: A discontinuous Galerkin method for the Cahn-Hilliard equation. J. Comput. Phys. **218**(2), 860–877 (2006). ISSN: 0021-9991. 10.1016/j.jcp.2006.03.010
100. Gómez, H., Calo, V.M., Bazilevs, Y., Hughes, T.J.: Isogeometric analysis of the Cahn-Hilliard phase-field model. Comput. Methods Appl. Mech. Eng. **197**(49–50), 4333–4352 (2008). 10.1016/j.cma.2008.05.003
101. Kästner, M., Metsch, P., De Borst, R.: Isogeometric analysis of the Cahn-Hilliard equation–a convergence study. J. Comput. Phys. **305**, 360–371 (2016). 10.1016/j.jcp.2015.10.047
102. Yang, Z., Lin, L., Dong, S.: A family of second-order energy-stable schemes for Cahn-Hilliard type equations. J. Comput. Phys. **383**, 24–54 (2019). 10.1016/j.jcp.2019.01.014
103. Zhang, R., Qian, X.: Triangulation-based isogeometric analysis of the Cahn–Hilliard phase-field model. Comput. Methods Appl. Mech. Eng. **357**, 112569 (2019). ISSN: 0045-7825. 10.1016/j.cma.2019.112569
104. Chan, C.L., Anitescu, C., Rabczuk, T.: Strong multipatch C1-coupling for isogeometric analysis on 2D and 3D domains. Comput. Methods Appl. Mech. Eng. **357**, 112599 (2019). 10.1016/j.cma.2019.112599
105. Copetti, M., Elliott, C.M.: Numerical analysis of the Cahn-Hilliard equation with a logarithmic free energy. Numer. Math. **63**(1), 39–65 (1992). 10.1007/BF01385847
106. Cherfils, L., Miranville, A., Zelik, S.: The Cahn-Hilliard equation with logarithmic potentials. Milan J. Math. **79**(2), 561–596 (2011). 10.1007/s00032-011-0165-4
107. Novick-Cohen, A., Segel, L.A.: Nonlinear aspects of the Cahn-Hilliard equation. Phys. D Nonlinear Phenomena **10**(3), 277–298 (1984). 10.1016/0167-2789(84)90180-5
108. Wodo, O., Ganapathysubramanian, B.: Computationally efficient solution to the Cahn-Hilliard equation: adaptive implicit time schemes, mesh sensitivity analysis and the 3D isoperimetric problem. J. Comput. Phys. **230**(15), 6037–6060 (2011). 10.1016/j.jcp.2011.04.012
109. Savina, T.V., Golovin, A.A., Davis, S.H., Nepomnyashchy, A.A., Voorhees, P.W.: Faceting of a growing crystal surface by surface diffusion. Phys. Rev. E **67**(2), 021606 (2003). 10.1103/PhysRevE.67.021606
110. Korzec, M., Evans, P.: From bell shapes to pyramids: A reduced continuum model for self-assembled quantum dot growth. Physica D Nonlinear Phenomena **239**(8), 465–474 (2010). 10.1016/j.physd.2010.01.014
111. Pawłow, I., Zajączkowski, W.M.: A sixth order Cahn-Hilliard type equation arising in oil-water-surfactant mixtures. Commun. Pure Appl. Anal. **10**(6), 1823 (2011). 10.3934/cpaa.2011.10.1823
112. Pawłow, I., Zajączkowski, W.M.: On a class of sixth order viscous Cahn-Hilliard type equations. Discrete Contin. Dyn. Syst. **6**(2), 517 (2013). 10.3934/dcdss.2013.6.517
113. Liu, C., Wang, J.: Some properties of solutions for a sixth-order Cahn-Hilliard type equation with inertial term. Appl. Anal. **97**(13), 2332–2348 (2018). 10.1080/00036811.2017.1364367

114. Du, Q., Ju, L., Li, X., Qiao, Z.: Stabilized linear semi-implicit schemes for the nonlocal Cahn–Hilliard equation. J. Comput. Phys. **363**, 39–54 (2018). ISSN: 0021-9991. 10.1016/j.jcp.2018.02.023
115. Yang, X., Zhao, J.: Efficient linear schemes for the nonlocal Cahn-Hilliard equation of phase field models. Comput. Phys. Commun. **235**, 234–245 (2019). 10.1016/j.cpc.2018.08.012
116. Korzec, M.D., Nayar, P., Rybka, P.: Global weak solutions to a sixth order Cahn-Hilliard type equation. SIAM J. Math. Anal. **44**(5), 3369–3387 (2012). 10.1137/100817590
117. Miranville, A.: Sixth-order Cahn-Hilliard systems with dynamic boundary conditions. Math. Methods Appl. Sci. **38**(6), 1127–1145 (2015). 10.1002/mma.3134
118. Miranville, A.: Sixth-order Cahn-Hilliard equations with singular nonlinear terms. Appl. Anal. **94**(10), 2133–2146 (2015). 10.1080/00036811.2014.972384
119. Boyarkin, O., Hoppe, R.H., Linsenmann, C.: High order approximations in space and time of a sixth order Cahn-Hilliard equation. Russ. J. Numer. Anal. Math. Model. **30**(6), 313–328 (2015). 10.1515/rnam-2015-0029

Chapter 4
Machine Learning Based Solutions of PDEs

4.1 Framework

In this chapter, we will focus on feed-forward deep neural network, also known as MLP. As the name suggests, the information in the network always travels forward (and not in cycles as RNN). Such networks have proven to be good for both regression and classification problems. The most important components of a feed-forward network are the weights, biases, nonlinear activation function and the back-propagation approach.

Consider a network of L layers, where $L > 3$. The first layer denotes the input layer and L-th layer is the output layer, while the network consists of $L-2$ hidden layers. The weighted input, z_i^l into i-th neuron on the layer, l, is a function of weight, W_{ij}^l and bias, b_j^l and is represented as:

$$z_i^l = \sigma_{l-1}\left(\sum_{j=1}^{m_{l-1}}\left(W_{i,j}^l(z_j^{l-1}) + b_i^l\right)\right), \tag{4.1}$$

where activation function in layer $l-1$, m_{l-1} are the number of neurons in the layer $l-1$, and $1 < l < (L-1)$. From the above concepts, the feed-forward algorithm for computing the output, Y^L is expressed as:

T. Rabczuk et al., *Scientific Machine Learning with Engineering Applications*,
Studies in Systems, Decision and Control 667,
https://doi.org/10.1007/978-3-032-20307-6_4

$$
\begin{aligned}
\boldsymbol{Y}^{L} &= \sigma_{L}(\mathbf{W}^{L+1}\boldsymbol{z}^{L} + \boldsymbol{b}^{L}), \\
\boldsymbol{z}^{L} &= \sigma_{L-1}\left(\mathbf{W}^{L}\boldsymbol{z}^{L-1} + \boldsymbol{b}^{L}\right), \\
\boldsymbol{z}^{L-1} &= \sigma_{L-2}\left(\mathbf{W}^{L-1}\boldsymbol{z}^{L-2} + \boldsymbol{b}^{L-1}\right), \\
&\vdots \\
\boldsymbol{z}^{1} &= \sigma_{0}\left(\mathbf{W}^{1}\boldsymbol{x} + \boldsymbol{b}^{1}\right),
\end{aligned}
\tag{4.2}
$$

where $\boldsymbol{x}$ is the input of the neural network in the first layer. Equation (4.2) can be represented in a compressed form as $Y = \mathcal{N}(x; \boldsymbol{\theta})$, where $\boldsymbol{\theta} = (\mathbf{W}, \boldsymbol{b})$ includes both the weights and biases of the neural network, $\mathcal{N}$. We must first learn the weights and biases of the neural network before we can use it. The initial step in this process is to identify a cost functional, $C(\mathcal{N}(x; \boldsymbol{\theta}))$ and requiring it to be reduced using discrete input points. To solve a one-dimensional differential equation of the type $-u''(x) = f(x)$, a cost function $C(\mathcal{N}(x; \boldsymbol{\theta})) = \mathcal{N}_{xx}(x; \boldsymbol{\theta}) + f(x)$ could be defined, where $\mathcal{N}_{xx}(x; \boldsymbol{\theta})$ is the second-order derivative with respect to x of the neural network. The higher-order derivatives can be easily computed by automatic differentiation technique as commonly implemented in existing machine learning frameworks. This functional can be minimized (in the sense of mean square) at a discrete set of *collocation points*.

Alternatively, we can define a cost function based on the Dirichlet minimization principle of the form, $\hat{C}(\mathcal{N}(x; \boldsymbol{\theta})) = \int_{\Omega}\left(\frac{1}{2}\left|\mathcal{N}_{x}(x; \boldsymbol{\theta})\right|^{2} - f(x)\mathcal{N}(x; \boldsymbol{\theta})\right) dx$, where Ω represents the computational domain and in this case, the integral can be computed by a suitable numerical quadrature scheme. The strategy for solving PDEs with computing the residual at the collocation points is referred to as the deep collocation method, while the approach to minimize the total energy computed at the quadrature points is commonly termed as deep energy method. In this section, we will cover both the approaches in details with examples to show their differences and similarity.

The collocation points or the quadrature points make up the *training data*, $\mathcal{D} = \{\boldsymbol{x}, \boldsymbol{Y}\}_{i=1}^{N_t}$, in the developed PINN approach, where we minimize the residual of the governing PDE or the variational energy of the system, respectively, written in terms of a loss-function. Common loss-functions used in the literature includes the l_1-loss function and the l_2-loss function, defined in (4.3).

$$
\begin{aligned}
l_1 &= \sum_{i=1}^{n} \left|Y_i - \mathcal{N}(x_i; \boldsymbol{\theta})\right|, \\
l_2 &= \sum_{i=1}^{n} \left(Y_i - \mathcal{N}(x_i; \boldsymbol{\theta})\right)^2,
\end{aligned}
\tag{4.3}
$$

where Y_i and $\mathcal{N}(x_i; \boldsymbol{\theta})$ are the target value and the corresponding predicted value, respectively and x_i denotes the sample point.

4.2 Physics-Informed Neural Networks

The goal of solving a boundary value problem using physics informed neural networks (PINNs) is to find a solution that satisfies the PDEs in its strong form at a set of collocation points. To this end, let us consider a general parametrized non-linear PDE of the form:

$$\begin{aligned} f(\boldsymbol{x}) &= \mathcal{D}[u](\boldsymbol{x}), \\ u(\boldsymbol{x}_d) &= u_d, \end{aligned} \tag{4.4}$$

defined on a bounded domain, where $\mathcal{D}$ is a differential operator, u_d is a scalar, and $u(\boldsymbol{x})$ is the unknown solution to be obtained at $\boldsymbol{x} = \{x_1, x_2, \dots x_n\}$ collocation points. The solution $u(\boldsymbol{x})$ needs to satisfy the boundary conditions at $x = x_d$. The latent solution of $u(\boldsymbol{x})$ can be approximated using the deep neural network with trainable parameters, $\boldsymbol{\theta}$, which is a collection of the weights and biases of the network. The cost function is the PDE residual, $r(\boldsymbol{x}, \boldsymbol{\theta})$ defined as:

$$r(\boldsymbol{x}, \boldsymbol{\theta}) = \mathcal{D}[u](\boldsymbol{x}) - f(\boldsymbol{x}). \tag{4.5}$$

As previously mentioned, the trainable parameters are learnt by minimizing the loss function, which is the sum of residual loss, $\mathcal{L}_r(\theta)$ and the boundary loss, $\mathcal{L}_d(\theta)$. The residual loss and the boundary loss could be defined as:

$$\begin{aligned} \mathcal{L}_r(\theta) &= \frac{1}{n}\sum_{i=1}^{n} |r(x^i, \theta)|, \\ \mathcal{L}_d(\theta) &= \frac{1}{n_d}\sum_{i=1}^{n_d} |u(x_d^i) - u_d|, \end{aligned} \tag{4.6}$$

where n_d are the points for the training data on which the Dirichlet boundary conditions are applied, while n are the total number of collocation points on which the residual loss is computed. In physics-informed neural networks, we finally minimize the loss, $\mathcal{L}(\theta) = \mathcal{L}_r(\theta) + \mathcal{L}_d(\theta)$ using suitable optimizers, such as those described in Sect. 2.3.

4.3 Deep Energy Method

In the deep energy method, the trainable parameters of the deep neural network are obtained by minimization of the total variational energy of the system. While the residual loss in Sect. 4.2 is computed at the collocation points, the variational loss is computed at the quadrature points (preferably). In this section, we will discuss in details the formulation of the deep energy method.

Let us consider a generic one dimensional time-independent differential equation of the form:

$$\mathcal{F}(u, u_x, \ldots, u_{x\cdots x}, x, f(x)) = 0, \tag{4.7a}$$

$$u(x_d) = u_d, \tag{4.7b}$$

where x_d represents a Dirichlet boundary point, $u(\boldsymbol{x})$ represents the dependent variable to be computed, and u_x, u_{xx} and $u_{x\cdots x}$ represents the derivatives of first, second and higher-orders with respect to the independent variable, $\boldsymbol{x}$. In (4.7a), $f(\boldsymbol{x})$ represents the source term. Let $\mathcal{V}_e$ represent the variational energy of (4.7a), and is defined as:

$$\mathcal{V}_e = \int_{\Omega} \mathcal{G}(u, u_x, \ldots, u_{x\cdots x}, x, f(x))\, d\Omega, \tag{4.8}$$

where $\mathcal{G}$ is a differentiable functional. With this, the solution to (4.7a) can be obtained by solving the following optimization problem:

$$\begin{aligned} u^* &= \operatorname{argmin}_u \mathcal{V}_e(u) \\ \text{subject to: } & u(x_D) = u_D. \end{aligned} \tag{4.9}$$

To solve this optimization problem, the loss function $\mathcal{L}(\theta)$ is the sum of the variational loss, $\mathcal{L}_v(\theta)$, and the boundary loss, $\mathcal{L}_d(\theta)$. The variational loss is defined as:

$$\mathcal{L}_v(\theta) = \frac{1}{n_g} \sum_{i=1}^{n_g} |\mathcal{V}_e(x_g^i)|, \tag{4.10}$$

where the loss is evaluated at the quadrature points, $\boldsymbol{x}_g$. It is important to note that the boundary loss represented either by the Dirichlet boundary and/or the non-homogeneous Neumann boundary is computed at uniformly distributed points over the boundary, hence the $\mathcal{L}_d(\theta)$ is the same as discussed in the previous section. Since the deep energy method is based on the energy principle (Galerkin's approach), the homogeneous Neumann boundary conditions are automatically satisfied, and hence not considered in the loss function.

Remark 4.1 The integral defined in (4.8), could also be computed at uniformly spaced points defined over the domain, however, a more accurate approach is to use the quadrature points. However, the quadrature points generated over the domain, are generally more dense near the boundary and sparse in the interiors. Hence, the results obtained will be inaccurate in case there exist local features where the solution is not smooth (e.g. in the presence of cracks). In the examples presented in this chapter, the problem domain is divided into elements, and the Gauss points are generated within each element.

Remark 4.2 In both the approaches, the boundary loss is satisfied by minimizing $\mathcal{L}_d(\theta)$. A penalty parameter often needs to be added with the boundary-loss term to establish a balance between the boundary-loss and the residual loss or the variational energy loss. This complicates the optimization problem because choosing appropriate penalty parameters is to modify the neural network outputs to ensure that the boundary criteria are met exactly, thereby eliminating the boundary loss term from the total loss function. To that end, we set,

$$\begin{aligned} u \approx u_{NN} &= \tilde{u}_D + B(x) \cdot \mathcal{N}(x; \boldsymbol{\theta}), \\ &= \hat{\mathcal{N}}(x; \boldsymbol{\theta}), \end{aligned} \tag{4.11}$$

where $\tilde{u}_D$ is a function chosen such that $\tilde{u}_D = u_D$ and $B(x) = 0$ on the Dirichlet boundary. As an example, if the boundary condition demands $u = 0$ at $x = 0$, we can set:

$$u \approx u_{NN} = x\mathcal{N}(x; \boldsymbol{\theta}). \tag{4.12}$$

The software for the Deep Energy Method is available at: ai4m.llmdem.com

4.4 Setting Up the Network Architecture

In the process of building the deep neural network, it is important to decide upon the following parameters that are essential to set up the deep neural network:

1. number of hidden layers,
2. number of neurons per layer,
3. activation function to be used in the hidden layers, and
4. the type and the number of optimizers.

There is only one input layer and one output layer in any neural network. The input layer's number of neurons is equal to the number of input variables in the data being processed, whereas the output layer's number of neurons is equal to the dimension of the output fields. According to the situation at hand, the number of hidden layers and neurons in each layer must be determined.

There is no specific rule for determining the number of hidden layers and neurons per hidden layer in general. Both of these characteristics have the potential to alter the network's final output. Under-fitting occurs when there are too few neurons in the hidden layers, making the network incapable of identifying signals in a complex data set. Over-fitting, on the other hand, occurs when there are too many neurons in the hidden layers with a fixed number of integration points, implying that the neural network has a richer approximation capacity than is required for the restricted amount of information in the training set. Additionally, as the number of hidden layers and neurons grows, so does the amount of time it takes to train the network. Obviously, a balance must be struck between too many and too few neurons, as well as the amount of hidden layers. Depending on the problem's complexity, these parameters are chosen through trial and error. In a linear elastic issue, for example, a smaller number of hidden layers and fewer neurons per layer are required to describe the approximation than in identifying complex non-linear interactions in coupled PDEs for fracture research.

The activation function transforms the input data in a non-linear way, allowing the neural network to simulate more complex correlations between input and output. The activation function must be differentiable in order for back-propagation to be possible. In general, there is no rule of thumb for determining which activation function is ideal in each circumstance. It's worth noting, though, that the output layer is always a linear function of the last concealed layer in order to recreate an unconstrained set of outputs.

The network parameters are calculated during the deep neural network's training to minimize the loss function. The speed of convergence is influenced by the optimization algorithm chosen. It is crucial to note, however, that finding a global optimum for non-convex optimization problems is not assured [1, 2]. The minimum reached using stochastic gradient descent algorithms is frequently reasonably close to the global optimum. Training a deep neural network entails optimizing all network's weights and biases. The number of parameters rises as the problem becomes more complex. Batch optimizers are the best option in these situations. The Adam optimizer is often the algorithm of choice among the available batch optimizers in *TensorFlow* due to its demonstrated performance. Adam uses adaptive step-sizes to achieve a type of learning rate annealing. Adam, on the other hand, does well in the first part of the training, where a rudimentary approximation of the optimal value can be obtained. We employ one of the quasi-Newton methods to find a local optimum around this zone.

4.5 Applications to Elliptic PDEs

This section focuses on the implementation of physics-informed neural networks (PINNs) and deep energy method for solving elliptic PDEs. The most common PDEs, which are also being considered here, are Poisson's equation, linear elasticity, and hyperelasticity. For elliptic problems, both the residual and energy based formulations can be used.

4.5.1 Poisson Equation

In the first example, we consider a two-dimensional Poisson's equation of the form

$$\begin{aligned} -\Delta u(x, y) &= f(x, y), \text{ for } (x, y) \in \Omega \\ f(x, y) &= -8\pi^2 \sin(2\pi x) \sin(2\pi y), \end{aligned} \tag{4.13}$$

defined in the domain $\Omega = [0, 1] \times [0, 1]$. We consider the Dirichlet boundary conditions: $u(x, y) = 0$ for $x \in \partial\Omega$. This problem is solved using both approaches, the physics informed neural networks (PINNs) and the deep energy method (DEM). To remove the influence of penalty terms, we multiply the output of the neural network by the function $B(x, y) = xy(1 - x)(1 - y)$, which enforces the Dirichlet boundary conditions exactly (cf. Remark 4.2).

In both formulations, the field approximation for $u(x, y)$ is represented by a neural network with two inputs neurons, corresponding to the values of x and y, and one output neuron, corresponding to the value of $u(x, y)$. The loss function is different for the two approaches, namely:

$$\mathcal{L}_{PINN}(\theta) = \frac{1}{N_{coll}} \sum_{l=1}^{N_{coll}} (\Delta u_\theta(x_i, y_i) - f(x_i, y_i))^2$$

$$\mathcal{L}_{DEM}(\theta) = \sum_{j=1}^{N_{int}} \left(\frac{1}{2} \left| \nabla u_\theta(x_j, y_j) \right|^2 - f(x_j, y_j) u(x_j, y_j) \right) w_j$$

Here θ represents the neural network trainable parameters (the weights and biases), N_{coll} the number of collocation points, N_{int} the number of integration points, w_j the corresponding integration weight for each quadrature point (x_j, y_j), and N_{bnd} is the number of boundary points.

We consider the PINN (Sect. 4.2) and DEM (Sect. 4.3) for solving this PDE. In the first case, we choose 5000 randomly distributed points in the interior of the domain. For the DEM, we consider a grid of 20×20 integration cells, with 4×4 Gauss integration points in each cell, for a total of 6400 integration points. The collocation and quadrature points are shown in Fig. 4.1. To speed-up the training process, we normalize the coordinates of the input points to the interval $[-1, 1]$, by applying the linear transformation $\hat{N}(X) = 2X - 1$. For both networks, we use three hidden layers, with 20 neurons in each layer, i.e. a [2, 20, 20, 20, 1] fully-connected NN. To train both networks, we first use the ADAM optimizer for 5000 iterations with a learning rate of 10^{-3}, followed by the BFGS optimizer for up to 5000 iterations using Hager-Zhang line search [3] as implemented in the Tensorflow-Probability library [4] (Fig. 4.2).

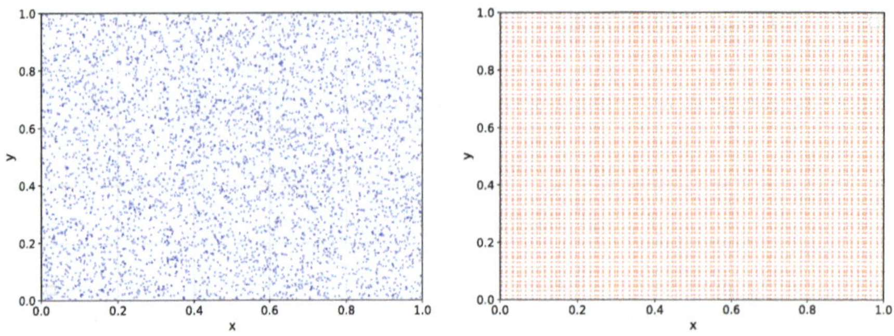

(a) Collocation Points (PINN) (b) Integration Points (DEM)

Fig. 4.1 The training points for PINN and DEM. (**a**) Collocation Points (PINN). (**b**) Integration Points (DEM)

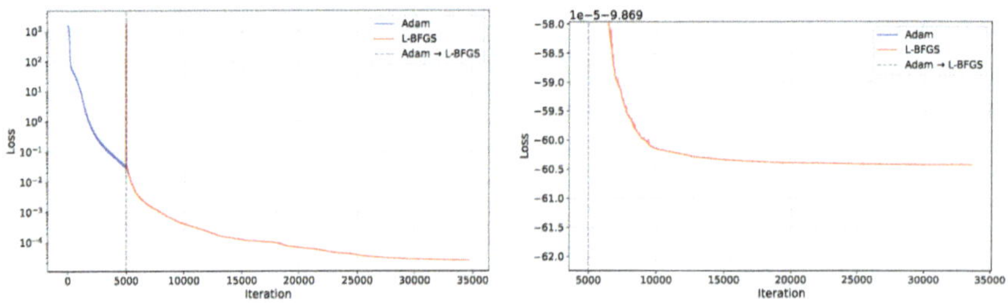

(a) PINN (b) DEM

Fig. 4.2 The convergence of the loss function for (**a**) PINN and (**b**) DEM for the Poisson equation example

The results for this example are shown in Fig. 4.3 together with the exact solution, which is in this case $u(x, y) = \sin(2\pi x)\sin(2\pi y)$. The convergence of the loss function with respect to the number of iterations for both methods is shown in Fig. 4.2. We note that the loss function for PINN approaches zero, while the loss function for DEM approaches the minimum energy of the PDE, which is:

$$\int_\Omega \left(\frac{1}{2}|\nabla u|^2(x, y) - u(x, y)f(x, y)\right) d\Omega = -\pi^2.$$

For this example, the results obtained by PINN are slightly more accurate than by DEM (relative L^2 error of $8 \cdot 10^{-6}$ vs. $2.5 \cdot 10^{-5}$), however the time taken by PINN is larger (137 seconds vs. 78 seconds on a Nvidia A4000 GPU) due to the additional cost of computing second order derivatives.

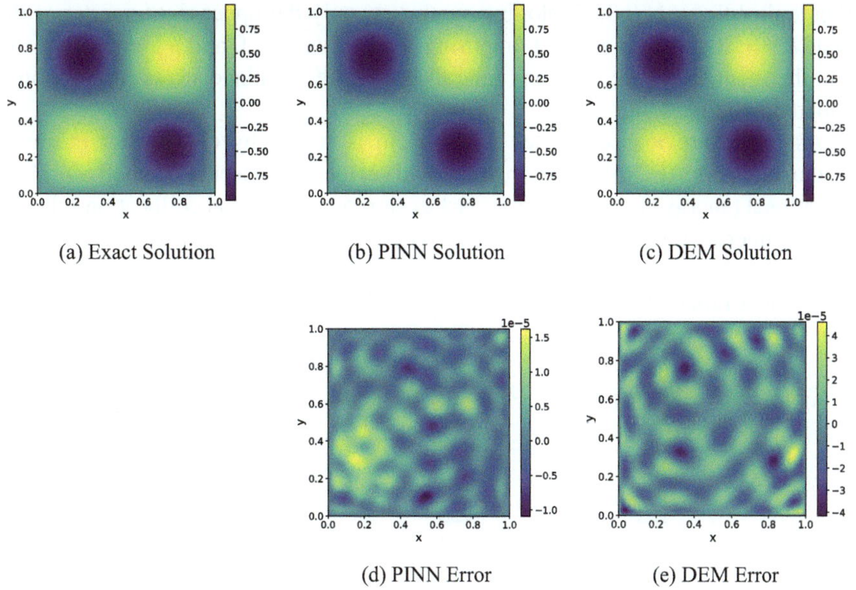

(a) Exact Solution (b) PINN Solution (c) DEM Solution

(d) PINN Error (e) DEM Error

Fig. 4.3 Comparison of exact solution, PINN solution, and DEM solution with their respective errors. (**a**) Exact solution. (**b**) PINN solution. (**c**) DEM solution. (**d**) PINN error. (**e**) DEM error

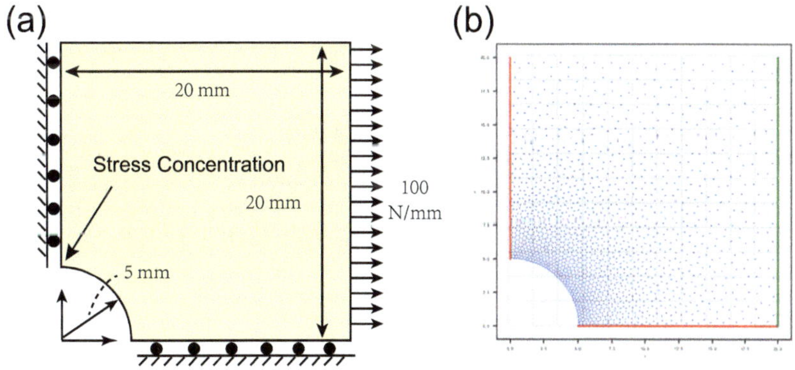

Fig. 4.4 Plate with a central hole (**a**) Geometry ($R = 5\,\text{mm}$, $L = 20\,\text{mm}$), with a uniform load of $100\,\text{N/mm}$ on the right edge. (**b**) Distribution of training points used in the DEM formulation

4.5.2 Linear Elasticity

A classical benchmark in linear elasticity is the square plate with a central circular hole, shown in Fig. 4.4. The original plate measures 40×40 mm and contains a hole of radius 5 mm. Both the left and right edges are subjected to a uniform traction of $t_x = 100\,\text{N/mm}$. The material parameters are Young's modulus $E = 1000\,\text{MPa}$ and

Poisson's ratio $\nu = 0.3$. Exploiting symmetry, only the upper-right quarter of the plate ($L = 20$ mm) is analysed, imposing $u_x = 0$ on $x = 0$ and $u_y = 0$ on $y = 0$; all other boundaries are traction (natural) boundaries. A plane-stress assumption is adopted; see Fig. 4.4a.

As in the previous example, the problem is solved by (1) Physics-Informed Neural Networks (PINNs) in strong form and (2) the Deep Energy Method (DEM) in variational (energy) form. Both approaches employ an identical collocation scheme: 75,298 interior points and 1000 points on each boundary, yielding five boundary subsets in total.

The loss function of the strong form of PINNs reads:

$$\begin{aligned}
\mathcal{L}_{\text{PINNs}} = \lambda_1 \sum_{i=1}^{N_{\text{pde}}} \Big[|\sigma_{xx,x}(\mathbf{x}_i) + \sigma_{xy,y}(\mathbf{x}_i)|^2 + |\sigma_{yx,x}(\mathbf{x}_i) + \sigma_{yy,y}(\mathbf{x}_i)|^2 \Big] \\
+ \lambda_2 \sum_{i=1}^{N_{\text{left}}} |u_x(\mathbf{x}_i)|^2 + \lambda_3 \sum_{i=1}^{N_{\text{left}}} |\sigma_{xy}(\mathbf{x}_i)|^2 \\
+ \lambda_4 \sum_{i=1}^{N_{\text{down}}} |u_y(\mathbf{x}_i)|^2 + \lambda_5 \sum_{i=1}^{N_{\text{down}}} |\sigma_{xy}(\mathbf{x}_i)|^2 \\
+ \lambda_6 \sum_{i=1}^{N_{\text{up}}} \Big[|\sigma_{yy}(\mathbf{x}_i)|^2 + |\sigma_{xy}(\mathbf{x}_i)|^2 \Big] \\
+ \lambda_7 \sum_{i=1}^{N_{\text{right}}} \Big[|\sigma_{xx}(\mathbf{x}_i) - t_x|^2 + |\sigma_{xy}(\mathbf{x}_i)|^2 \Big] \\
+ \lambda_8 \sum_{i=1}^{N_{\text{circle}}} \Big[|\sigma_{xx} n_x + \sigma_{xy} n_y|^2 + |\sigma_{yx} n_x + \sigma_{yy} n_y|^2 \Big],
\end{aligned} \tag{4.14}$$

with the constitutive and kinematic relations

$$\sigma_{ij} = \frac{E}{1+\nu}\,\varepsilon_{ij} + \frac{E\nu}{(1+\nu)(1-2\nu)}\,\varepsilon_{kk}\,\delta_{ij}, \qquad \varepsilon_{ij} = \tfrac{1}{2}\big(u_{i,j} + u_{j,i}\big). \tag{4.15}$$

$\{\lambda_i\}_{i=1}^{8} = \{10,\ 2000,\ 1,\ 1,\ 2000,\ 1,\ 1,\ 1\}$ are selected empirically owing to the large number of hyper-parameters. It is a challenge to get the optimal hyperparameters in the strong form of PINNs [5–7].

The displacement field is normalised as

$$\mathbf{u}(\mathbf{x}) \approx \mathbf{u}\big(\tfrac{\mathbf{x}}{L};\ \boldsymbol{\theta}\big), \tag{4.16}$$

where $\boldsymbol{\theta}$ denotes the network parameters.

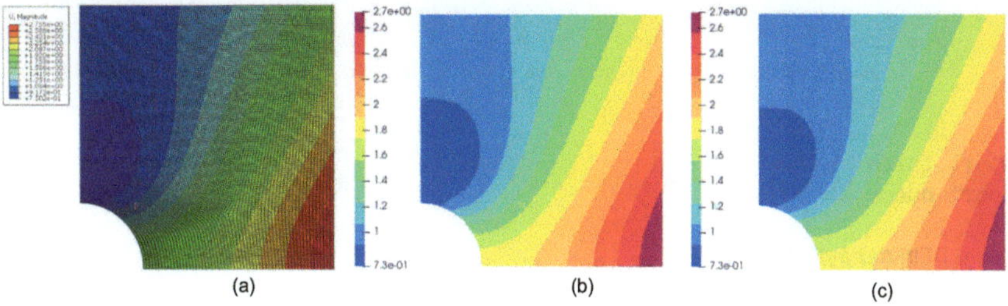

Fig. 4.5 Predicted displacement magnitude (**a**) FEM reference; (**b**) PINNs; (**c**) DEM

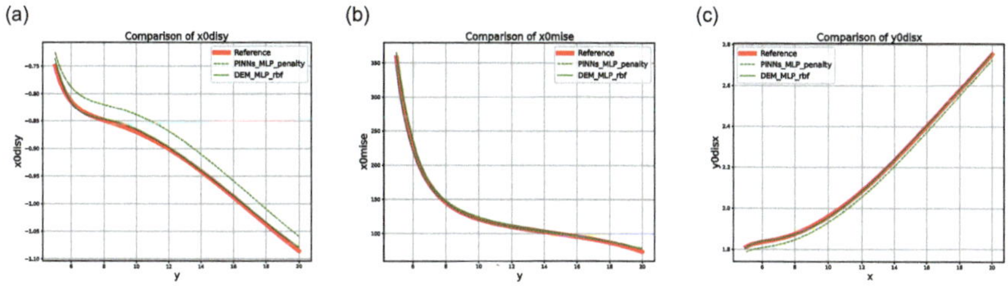

Fig. 4.6 Line comparison along symmetry axes (**a**) u_y along $x = 0$; (**b**) Mises stress along $x = 0$; (**c**) u_x along $y = 0$

Table 4.1 Accuracy and efficiency for the plate with a central hole

	Relative error			
Method	$\mathcal{L}_2$ (disp.)	$\mathcal{H}_1$ (Mises)	NN architecture	Time (s/1000 epochs)
PINNs	0.00858	0.00649	[2, 30, 30, 30, 30, 2]	70.65
DEM	0.00154	0.00115	[2, 30, 30, 30, 30, 2]	14.57

The loss function of the energy form of PINNs (DEM) is

$$\mathcal{L}_{\text{DEM}} = \int_{\Omega} \Psi \, \mathrm{d}\Omega - \int_{\Gamma_{\text{right}}} t_x \, u_x \, \mathrm{d}\Gamma, \qquad \Psi = \tfrac{1}{2}\sigma_{ij}\varepsilon_{ij}, \tag{4.17}$$

with admissible displacements enforced a priori through $u_x = x\,\hat{u}_x(\mathbf{x}/L;\boldsymbol{\theta}),\ u_y = y\,\hat{u}_y(\mathbf{x}/L;\boldsymbol{\theta})$. Triangular numerical integration is adopted, which has proven more accurate than Monte-Carlo sampling for DEM.

Figure 4.5 compares the predicted displacement magnitude $u_{\text{mag}} = \sqrt{u_x^2 + u_y^2}$ obtained from FEM (reference), PINNs, and DEM. The FEM model uses 15,300 CPS8R elements and yields a converged solution in 30.64 s.

Because of the stress concentration at the hole rim, FEM typically requires local mesh refinement. The maximum u_y and Mises stress occur along $x = 0$, whereas the maximum u_x occurs along $y = 0$. Figure 4.6 presents these line plots, demonstrating that DEM matches the reference solution more closely than PINNs.

Table 4.1 summarises accuracy and runtime. DEM outperforms PINNs in both accuracy and computational cost for this linear-elastic problem. The workflow of the DEM implementation is listed in Listing 10.

```python
"""
DEM (Deep Energy Method) Demo - Energy Construction
Purpose: Demonstrate energy functional construction for elasticity problems
"""

import torch
import numpy as np

class DEMEnergyDemo:
    def strain_energy_density(self, u, x):
        """Calculate strain energy density"""
        # Compute strain gradients
        dux_dx = torch.autograd.grad(u[:, 0], x, torch.ones_like(u[:, 0]),
                                     create_graph=True)[0][:, 0]
        duy_dy = torch.autograd.grad(u[:, 1], x, torch.ones_like(u[:, 1]),
                                     create_graph=True)[0][:, 1]
        dux_dy = torch.autograd.grad(u[:, 0], x, torch.ones_like(u[:, 0]),
                                     create_graph=True)[0][:, 1]
        duy_dx = torch.autograd.grad(u[:, 1], x, torch.ones_like(u[:, 1]),
                                     create_graph=True)[0][:, 0]
        # Strain components
        exx = dux_dx
        eyy = duy_dy
        exy = 0.5 * (dux_dy + duy_dx)
        # Strain energy density (linear elasticity)
        energy_density = 0.5 * (self.D11 * exx**2 + self.D11 * eyy**2 +
                                2*self.D12 * exx * eyy + 2*self.D33 * exy**2)
        return energy_density

    def external_work(self, u_boundary, force_boundary):
        """Calculate external work on boundary"""
        # External work = force * displacement
        work_density = torch.sum(u_boundary * force_boundary, dim=1)
        return work_density

    def total_energy(self, x_domain, x_boundary, force_boundary, model):
        """Calculate total energy functional"""
        # Internal energy (strain energy)
        u_domain = self.get_displacement(x_domain, model)
        strain_energy = self.strain_energy_density(u_domain, x_domain)
        internal_energy = torch.sum(strain_energy)  # Simple summation
        # External work
        u_boundary = self.get_displacement(x_boundary, model)
        external_work = self.external_work(u_boundary, force_boundary)
        total_external_work = torch.sum(external_work)
        # Total energy = Internal energy - External work
        total_energy = internal_energy - total_external_work
        return total_energy
```

Listing 10: Python demo for DEM in linear elasticity

4.5.3 Hyperelasticity

This example demonstrates the effectiveness of scientific machine learning in solving nonlinear hyperelasticity problems. Such problems are nonlinear in both the constitutive law and the geometric kinematics. We consider the Neo-Hookean model, where the strain energy density function is defined as:

$$
\begin{aligned}
\Psi &= \frac{1}{2}\lambda(\ln J)^2 - \mu \ln J + \frac{1}{2}\mu(I_1 - 3), \\
J &= \det(\boldsymbol{F}), \\
I_1 &= \mathrm{trace}(\boldsymbol{C}), \\
\boldsymbol{C} &= \boldsymbol{F}^T \boldsymbol{F},
\end{aligned}
\tag{4.18}
$$

where Ψ is the strain energy, and λ and μ are the Lamé parameters:

$$
\begin{cases}
\lambda = \frac{\nu E}{(1+\nu)(1-2\nu)}, \\
\mu = \frac{E}{2(1+\nu)},
\end{cases}
\tag{4.19}
$$

with E being the Young's modulus and ν the Poisson's ratio. $\boldsymbol{F}$ is the deformation gradient:

$$
\boldsymbol{F} = \frac{\partial \boldsymbol{x}}{\partial \boldsymbol{X}},
\tag{4.20}
$$

where $\boldsymbol{x} = \boldsymbol{X} + \boldsymbol{u}$ is the spatial coordinate and $\boldsymbol{X}$ is the material coordinate.

Since the problem admits a variational structure, we solve it using the Deep Energy Method (DEM). Solving this problem with PINNs in the strong form is cumbersome and less efficient due to the higher-order derivatives involved. For detailed justification, refer to [8]. Thus, we only consider the energy-based DEM formulation, whose loss function is given by:

$$
\mathcal{L} = \int_{\Omega} (\Psi - \boldsymbol{f} \cdot \boldsymbol{u})\, d\Omega - \int_{\Gamma^t} \bar{\boldsymbol{t}} \cdot \boldsymbol{u}\, d\Gamma,
\tag{4.21}
$$

where $\boldsymbol{f}$ is the body force, and $\bar{\boldsymbol{t}}$ is the surface traction on the boundary Γ^t.

We use a fully connected MLP to solve a benchmark 2D cantilever beam problem under plane stress, illustrated in Fig. 4.7. Since strong-form PINNs and BINNs are not suitable for this problem, we restrict ourselves to DEM.

We test several numerical integration strategies for DEM, including Monte Carlo, trapezoidal rule, and Simpson's rule. A uniform grid with 200 points along the length and 50 along the height results in 10,000 domain points.

We use the L-BFGS optimizer. Table 4.2 compares the accuracy and runtime for the three integration methods. FEM is used as the reference. Due to singularities near

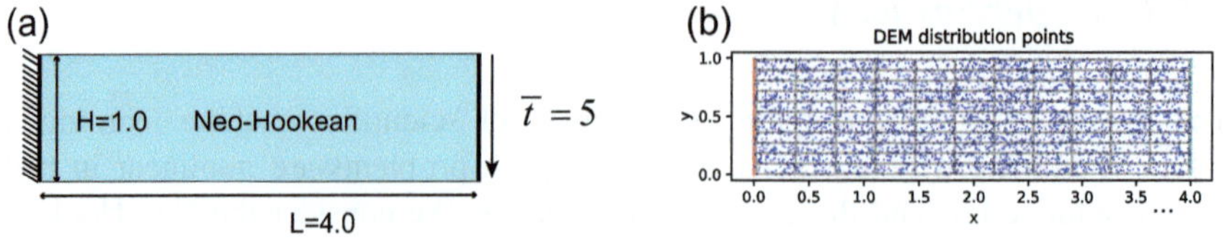

Fig. 4.7 Setup of the Neo-Hookean hyperelastic cantilever beam. (**a**) Geometry: height $H = 1.0$, length $L = 4.0$, with $E = 1000$ and $\nu = 0.3$. The left boundary ($x = 0$) is clamped, and a uniform downward traction $\bar{t} = 5$ is applied on the right. (**b**) DEM point distribution: blue (domain), red (essential BC), green (traction BC)

Table 4.2 Comparison of DEM performance with different integration methods for the hyperelasticity problem

Algorithm	Displacement relative error ($\mathcal{L}_2$)	NN architecture	Parameters	Time (s, 100 Epochs)
DEM_Mont	0.04756	[2, 30,30,30,30,2]	2942	37.34
DEM_Trap	0.01524	[2, 30,30,30,30,2]	2942	35.76
DEM_Simp	0.01753	[2, 30,30,30,30,2]	2942	39.87

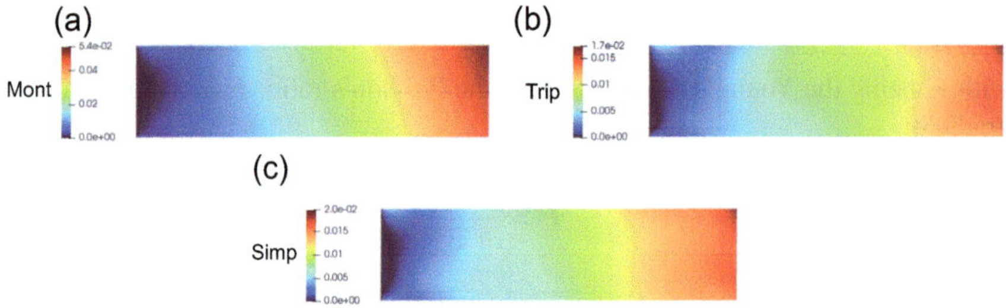

Fig. 4.8 Absolute error contours of vertical displacement u_y: (**a**) Monte Carlo, (**b**) trapezoidal, and (**c**) Simpson's integration

the corners (0, 0) and (0, 1), only the L_2 error of displacement is considered. Monte Carlo integration results in larger errors, while the trapezoidal and Simpson's rule yield comparable accuracy, with Simpson's rule offering no significant improvement despite its higher-order accuracy.

The vertical displacement u_y is the primary response in the cantilever beam under vertical load. Figure 4.8 shows the absolute error contours of u_y for each integration method. Figure 4.9 tracks the evolution of the L_2 error in the total displacement $u_{\text{mag}} = \sqrt{u_x^2 + u_y^2}$.

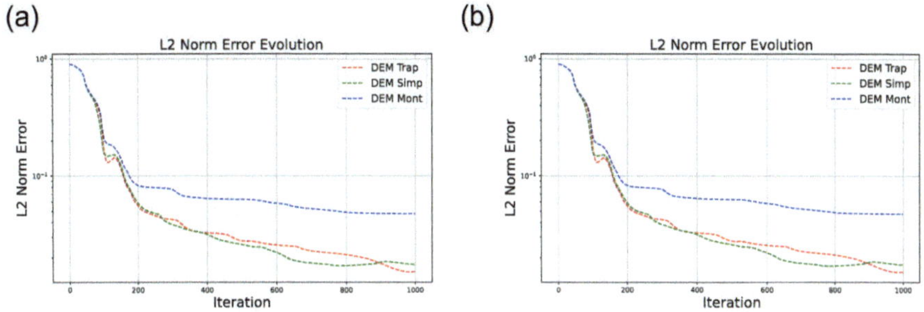

Fig. 4.9 Evolution of the L_2 relative error of displacement u_{mag} during training

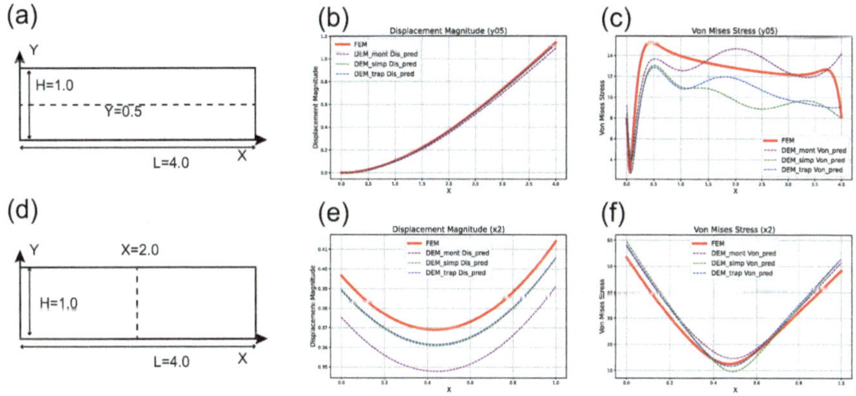

Fig. 4.10 Line plots of absolute displacement and Von Mises stress for the Neo-Hookean cantilever beam. Top row: (**b**) displacement and (**c**) stress along $y = 0.5$ (**a**). Bottom row: (**e**) displacement and (**f**) stress along $x = 2.0$ (**d**)

Despite corner singularities, we compare absolute displacement and Von Mises stress along $x = 2$ and $y = 0.5$. Figure 4.10 shows the predictions using DEM with the three integration strategies.

Listing 11 presents the Python implementation of DEM for hyperelasticity.

4.6 Applications to Parabolic PDEs

4.6.1 Heat Equation

The heat equation is a fundamental partial differential equation that describes the distribution of heat (or temperature variation) in a given region over time. For a

```python
"""
DEM (Deep Energy Method) Demo - Hyperelasticity
Purpose: Demonstrate energy functional construction for hyperelastic problems
"""
class DEMHyperelasticDemo:
    def deformation_gradient(self, u, x):
        """Calculate deformation gradient F = I + u"""
        # Compute displacement gradients
        dux_dx = torch.autograd.grad(u[:, 0], x, torch.ones_like(u[:, 0]),
                                     create_graph=True)[0][:, 0]
        duy_dy = torch.autograd.grad(u[:, 1], x, torch.ones_like(u[:, 1]),
                                     create_graph=True)[0][:, 1]
        dux_dy = torch.autograd.grad(u[:, 0], x, torch.ones_like(u[:, 0]),
                                     create_graph=True)[0][:, 1]
        duy_dx = torch.autograd.grad(u[:, 1], x, torch.ones_like(u[:, 1]),
                                     create_graph=True)[0][:, 0]

        # Deformation gradient components
        F11 = 1.0 + dux_dx
        F12 = dux_dy
        F21 = duy_dx
        F22 = 1.0 + duy_dy
        return F11, F12, F21, F22
    def neo_hookean_energy_density(self, u, x):
        """Calculate Neo-Hookean strain energy density"""
        F11, F12, F21, F22 = self.deformation_gradient(u, x)
        # Determinant of F
        J = F11 * F22 - F12 * F21
        # First invariant of C = F^T F
        Ic = F11**2 + F12**2 + F21**2 + F22**2
        # Neo-Hookean strain energy density
        energy_density = 0.5 * self.mu * (Ic - 2) - self.mu * torch.log(J) + \
                        0.5 * self.lmbda * (torch.log(J))**2
        return energy_density

    def external_work(self, u_boundary, force_boundary):
        """Calculate external work on boundary"""
        # External work = force * displacement
        work_density = torch.sum(u_boundary * force_boundary, dim=1)
        return work_density

    def total_energy(self, x_domain, x_boundary, force_boundary, model):
        """Calculate total energy functional"""
        # Internal energy (strain energy)
        u_domain = self.get_displacement(x_domain, model)
        strain_energy = self.neo_hookean_energy_density(u_domain, x_domain)
        internal_energy = torch.sum(strain_energy)  # Simple summation
        # External work
        u_boundary = self.get_displacement(x_boundary, model)
        external_work = self.external_work(u_boundary, force_boundary)
        total_external_work = torch.sum(external_work)
        # Total energy = Internal energy - External work
        total_energy = internal_energy - total_external_work
        return total_energy
```

Listing 11: Python code for solving hyperelasticity problems using DEM two-dimensional domain $\Omega \subset \mathbb{R}^2$ with boundary $\partial\Omega$, the heat equation is given by:

$$\frac{\partial u}{\partial t} = \alpha \Delta u = \alpha \left(\frac{\partial^2 u}{\partial x^2} + \frac{\partial^2 u}{\partial y^2} \right) \tag{4.22}$$

where $u(x, y, t)$ represents the temperature distribution, α is the thermal diffusivity constant, and Δ is the Laplacian operator. The equation is subject to initial and boundary conditions:

$$u(x, y, 0) = u_0(x, y), \quad \text{for } (x, y) \in \Omega \tag{4.23}$$

$$u(x, y, t) = g(x, y, t), \quad \text{for } (x, y) \in \partial\Omega,\ t > 0 \tag{4.24}$$

where $u_0(x, y)$ is the initial temperature distribution and $g(x, y, t)$ specifies the boundary conditions.

The PINN approach for solving the heat equation involves training a neural network $u_\theta(x, y, t)$ to approximate the solution by minimizing a composite loss function that enforces the physics constraints. The total loss function $\mathcal{L}_{total}$ consists of three components:

$$\mathcal{L}_{total} = \lambda_{PDE}\mathcal{L}_{PDE} + \lambda_{IC}\mathcal{L}_{IC} + \lambda_{BC}\mathcal{L}_{BC} \tag{4.25}$$

Here, $\mathcal{L}_{PDE}$ is the PDE residual loss, $\mathcal{L}_{IC}$ is the initial condition loss, and $\mathcal{L}_{BC}$ is the boundary condition loss, and λ_{PDE}, λ_{IC}, and λ_{BC} are penalty weights applied to each of the terms. PDE residual loss ensures that the neural network solution satisfies the heat equation at sampled collocation points within the domain:

$$\mathcal{L}_{PDE} = \frac{1}{N_{PDE}} \sum_{i=1}^{N_{PDE}} \left| \frac{\partial u_\theta}{\partial t}(x_i, y_i, t_i) - \alpha \left(\frac{\partial^2 u_\theta}{\partial x^2}(x_i, y_i, t_i) + \frac{\partial^2 u_\theta}{\partial y^2}(x_i, y_i, t_i) \right) \right|^2 \tag{4.26}$$

where (x_i, y_i, t_i) are randomly sampled collocation points in the spatiotemporal domain.

The initial condition loss enforces the prescribed initial temperature distribution:

$$\mathcal{L}_{IC} = \frac{1}{N_{IC}} \sum_{i=1}^{N_{IC}} |u_\theta(x_i, y_i, 0) - u_0(x_i, y_i)|^2 \tag{4.27}$$

The boundary condition loss ensures that the solution satisfies the prescribed boundary conditions:

$$\mathcal{L}_{BC} = \frac{1}{N_{BC}} \sum_{i=1}^{N_{BC}} |u_\theta(x_i, y_i, t_i) - g(x_i, y_i, t_i)|^2 \tag{4.28}$$

where (x_i, y_i, t_i) are points sampled on the boundary $\partial\Omega \times [0, T]$.

To verify our PINN implementation, we consider a benchmark problem with a known analytical solution. We solve the heat equation on the unit square $\Omega = [0, 1] \times [0, 1]$ with $\alpha = 0.1$ and the following manufactured solution:

$$u_{exact}(x, y, t) = e^{-2\pi^2 \alpha t} \sin(\pi x) \sin(\pi y) \tag{4.29}$$

This solution satisfies the heat equation with homogeneous Dirichlet boundary conditions ($u = 0$ on $\partial\Omega$) and the initial condition:

$$u_0(x, y) = \sin(\pi x) \sin(\pi y) \tag{4.30}$$

The PINN is implemented using a fully connected neural network with the following architecture:

- Input layer: 3 neurons (x, y, t)
- Hidden layers: 4 layers with 20 neurons each
- Activation function: swish ($\sigma(x) = \frac{x}{1+\exp(-x)}$)
- Output layer: 1 neuron (temperature u)

The network is trained using a two-phase optimization strategy:

1. **Adam optimizer**: 2000 iterations with learning rate 10^{-3} for initial training
2. **L-BFGS optimizer**: 1000 iterations using TensorFlow Probability's BFGS implementation for refinement

We sample 2500 collocation points for the PDE residual, 1000 points for the initial condition, and 1000 total boundary points (250 per edge). The implementation uses JAX with 64-bit floating-point precision.

Figure 4.11 shows the PINN solution compared to the exact solution at different time snapshots. The network successfully captures the heat diffusion process, with the temperature decreasing over time due to the homogeneous boundary conditions (Fig. 4.12).

Figure 4.13 shows both the training points distribution and loss convergence history. Figure 4.13a shows the distribution of training points. The PDE collocation points are distributed throughout the domain (excluding $t = 0$ to avoid conflicts with the initial condition), while boundary points are uniformly sampled along all four edges at various times. Figure 4.13b illustrates the training loss history for both optimization phases. The Adam optimizer rapidly reduces the loss during the initial 2000 iterations, while the L-BFGS optimizer provides further refinement, achieving a final loss value of approximately 10^{-5}. Note that the L-BFGS training loss includes the values obtained during the line search process. Therefore, there are close to 12,000 loss function evaluation values being plotted, although only 2000 L-BFGES iterations have been specified.

The results show that PINNs can solve the heat equation with good accuracy. By combining the Adam and L-BFGS optimizers, along with the swish activation

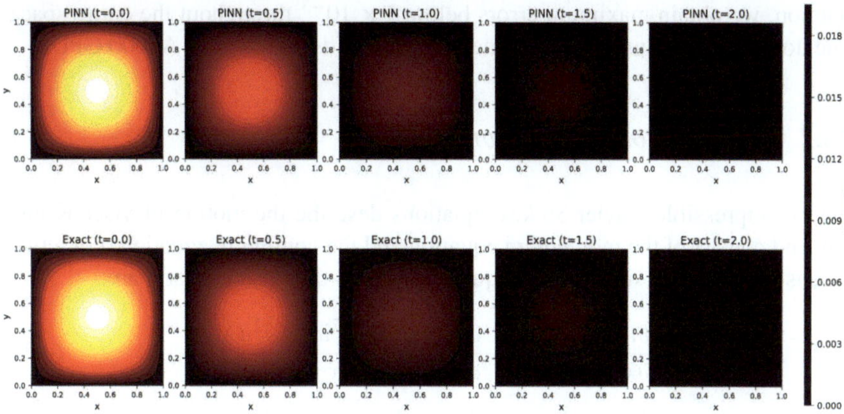

Fig. 4.11 PINN solution of the 2D heat equation at different time instances. Top row: PINN predictions. Bottom row: Exact solution. The colorbar indicates temperature values

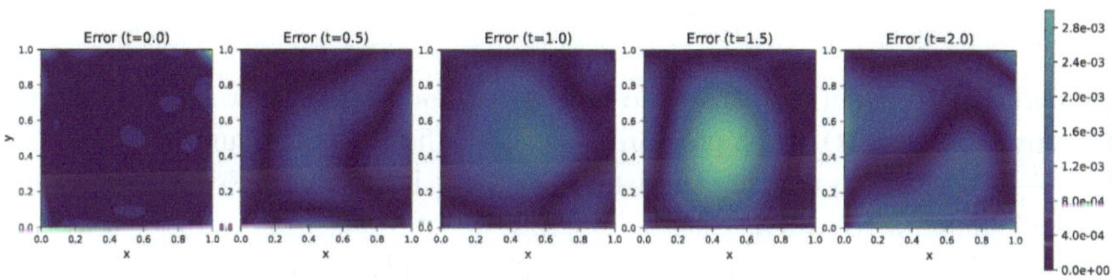

Fig. 4.12 Absolute error between PINN predictions and exact solution at different time instances

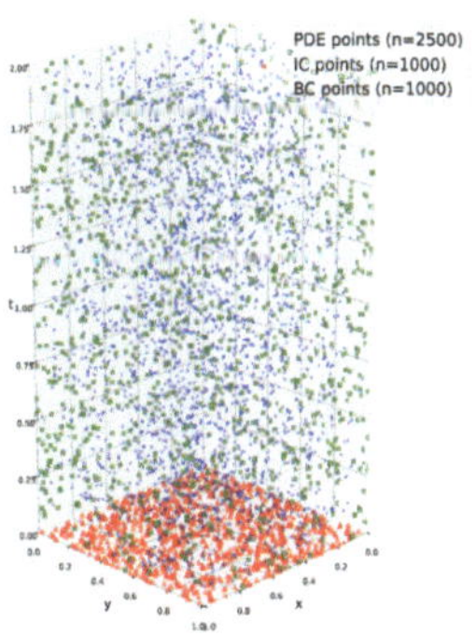

(a) Training points distribution

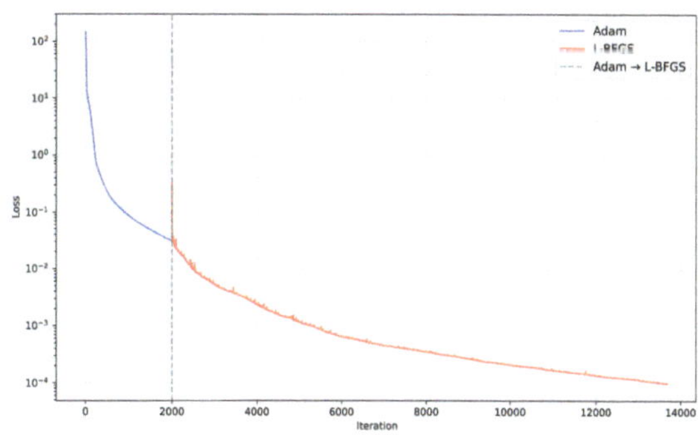

(b) Loss convergence history during training.

Fig. 4.13 Training visualization: (**a**) 3D distribution of training points showing blue circles for PDE collocation points (n = 2500), red triangles for initial condition points at $t = 0$ (n = 1000), and green squares for boundary condition points (n = 400) with aspect ratio reflecting domain dimensions $T = 2L$, and (**b**) training loss history showing the transition from Adam to L-BFGS optimization with the vertical dashed line indicating the switch between optimizers. (**a**) Training points distribution. (**b**) Loss convergence history during training function, we obtain maximum errors below 3×10^{-3} throughout the entire space-time domain. The computational time is 209 seconds on a Nvidia A4000 GPU.

4.6.2 Navier-Stokes Equation

The incompressible Navier-Stokes equations describe the motion of viscous fluid flow and consist of the momentum equation and the continuity equation. For a two-dimensional domain $\Omega \subset \mathbb{R}^2$, the equations in non-dimensional form are:

$$\frac{\partial u}{\partial t} + u\frac{\partial u}{\partial x} + v\frac{\partial u}{\partial y} = -\frac{\partial p}{\partial x} + \frac{1}{Re}\left(\frac{\partial^2 u}{\partial x^2} + \frac{\partial^2 u}{\partial y^2}\right) \tag{4.31}$$

$$\frac{\partial v}{\partial t} + u\frac{\partial v}{\partial x} + v\frac{\partial v}{\partial y} = -\frac{\partial p}{\partial y} + \frac{1}{Re}\left(\frac{\partial^2 v}{\partial x^2} + \frac{\partial^2 v}{\partial y^2}\right) \tag{4.32}$$

$$\frac{\partial u}{\partial x} + \frac{\partial v}{\partial y} = 0 \tag{4.33}$$

where $u(x, y, t)$ and $v(x, y, t)$ are the velocity components in the x and y directions respectively, $p(x, y, t)$ is the pressure, and Re is the Reynolds number defined as $Re = \frac{UL}{\nu}$, where U is the characteristic velocity, L is the characteristic length, and ν is the kinematic viscosity.

For the Navier-Stokes equations, we employ a fully connected neural network to approximate the velocity components and pressure: $u_\theta(x, y, t)$, $v_\theta(x, y, t)$, and $p_\theta(x, y, t)$. The total loss function consists of several components:

$$\mathcal{L}_{total} = \mathcal{L}_{NS} + \mathcal{L}_{IC} + \mathcal{L}_{BC} + \lambda_p \mathcal{L}_p \tag{4.34}$$

The Navier-Stokes residual loss enforces the governing equations at collocation points:

$$\mathcal{L}_{NS} = \frac{1}{N_{NS}} \sum_{i=1}^{N_{NS}} \left(|r_u^i|^2 + |r_v^i|^2 + |r_{cont}^i|^2\right) \tag{4.35}$$

where the residuals are:

$$r_u^i = \frac{\partial u_\theta}{\partial t} + u_\theta\frac{\partial u_\theta}{\partial x} + v_\theta\frac{\partial u_\theta}{\partial y} + \frac{\partial p_\theta}{\partial x} - \frac{1}{Re}\left(\frac{\partial^2 u_\theta}{\partial x^2} + \frac{\partial^2 u_\theta}{\partial y^2}\right) \tag{4.36}$$

$$r_v^i = \frac{\partial v_\theta}{\partial t} + u_\theta\frac{\partial v_\theta}{\partial x} + v_\theta\frac{\partial v_\theta}{\partial y} + \frac{\partial p_\theta}{\partial y} - \frac{1}{Re}\left(\frac{\partial^2 v_\theta}{\partial x^2} + \frac{\partial^2 v_\theta}{\partial y^2}\right) \tag{4.37}$$

$$r_{cont}^i = \frac{\partial u_\theta}{\partial x} + \frac{\partial v_\theta}{\partial y} \tag{4.38}$$

evaluated at collocation points (x_i, y_i, t_i).

The initial condition loss enforces the prescribed initial velocity field:

$$\mathcal{L}_{IC} = \frac{1}{N_{IC}} \sum_{i=1}^{N_{IC}} \left(|u_\theta(x_i, y_i, 0) - u_0(x_i, y_i)|^2 + |v_\theta(x_i, y_i, 0) - v_0(x_i, y_i)|^2 \right) \tag{4.39}$$

The boundary condition loss enforces no-slip or prescribed velocity conditions:

$$\mathcal{L}_{BC} = \frac{1}{N_{BC}} \sum_{i=1}^{N_{BC}} \left(|u_\theta(x_i, y_i, t_i) - u_{bc}(x_i, y_i, t_i)|^2 + |v_\theta(x_i, y_i, t_i) - v_{bc}(x_i, y_i, t_i)|^2 \right) \tag{4.40}$$

Since pressure is determined up to a constant in incompressible flow, we include a pressure regularization term:

$$\mathcal{L}_p = \left(\frac{1}{N_p} \sum_{i=1}^{N_p} p_\theta(x_i, y_i, t_i) \right)^2 \tag{4.41}$$

This ensures the mean pressure remains close to zero, preventing drift during training.

4.6.2.1 Benchmark Example 1: Lid-Driven Cavity Flow

The lid-driven cavity flow is a classic benchmark problem for validating numerical methods. We consider a square cavity $\Omega = [0, 1] \times [0, 1]$ with the following boundary conditions:

$$u(x, 1, t) = 1, \quad v(x, 1, t) = 0 \quad \text{(top lid)} \tag{4.42}$$

$$u = v = 0 \quad \text{(other walls)} \tag{4.43}$$

For steady-state analysis, we set $\frac{\partial}{\partial t} = 0$ and solve at Reynolds number $Re = 100$.

The PINN architecture employs a single unified network for velocity components (u, v) and pressure p. The network consists of 4 hidden layers with 32 neurons each, using tanh activation functions for the hidden layers. Input normalization to the range $[-1, 1]$ is applied to improve training stability. As in the previous examples, the implementation uses JAX with 64-bit precision for better numerical accuracy.

The training strategy employs a two-phase optimization approach using TensorFlow Probability's L-BFGS optimizer. The training data consists of 1600 collocation points uniformly distributed within the domain and 400 boundary points (100 per edge) to enforce the boundary conditions. The first phase uses the Adam optimizer for 5000 iterations with a learning rate of 10^{-3} to provide a good initial guess. The second phase switches to the L-BFGS optimizer for 5000 iterations to achieve fine convergence. Enhanced loss weighting is applied with $\mathcal{L}_{total} = \mathcal{L}_{NS} + 100\mathcal{L}_{BC} + 0.01\mathcal{L}_p$ to ensure proper enforcement of boundary conditions while maintaining numerical stability through pressure regularization. The total training time is approximately 8 minutes on standard hardware.

The PINN solution for the lid-driven cavity flow at $Re = 100$ was obtained using a two-phase optimization strategy. The Adam optimizer ran for 77 seconds, followed by L-BFGS optimization which completed in 402 seconds with 5000 iterations. The total training time was 479 seconds (approximately 8 minutes) on an Nvidia A4000 GPU.

The final loss achieved was 4.673×10^{-5}, showing that the governing equations are satisfied with fairly small error. The accuracy of the solution was validated against the benchmark data from [9]. The U-velocity error was measured at 0.63%, while the V-velocity error was 1.90%, resulting in an average error of 1.26%. These low error values confirm that the PINN approach accurately captures the steady-state flow features of the lid-driven cavity problem.

Figure 4.14 presents the key aspects of the PINN solution for the lid-driven cavity flow. Panel (a) shows the distribution of collocation points, with interior points for enforcing the PDE residuals and boundary points concentrated along the cavity walls for enforcing no-slip conditions and the lid velocity. The placement of these points ensures adequate resolution of both the bulk flow and boundary layers. Panel (b) displays the velocity magnitude contours, revealing the characteristic high-velocity region near the moving lid and the gradual decrease in velocity toward the cavity center and bottom corners. The maximum velocity occurs near the lid, as expected from the driven flow physics. Panel (c) illustrates the streamlines, clearly showing the primary vortex centered at approximately (0.617, 0.734), which agrees well with benchmark data. Secondary vortices in the bottom corners, typical of lid-driven cavity flows at this Reynolds number, are also captured. Panel (d) demonstrates the loss convergence history during training, showing the rapid initial decrease during the Adam optimization phase followed by fine-tuning during the L-BFGS phase. The transition between optimizers is clearly visible, with L-BFGS providing the final convergence to machine precision levels.

4.6.2.2 Benchmark Example 2: Taylor-Green Vortex

The Taylor-Green vortex is an analytical solution to the 2D Navier-Stokes equations that describes the decay of vortices due to viscous effects. The initial conditions on domain $\Omega = [0, 2\pi] \times [0, 2\pi]$ with periodic boundary conditions are:

$$u(x, y, 0) = \sin(x)\cos(y) \tag{4.44}$$

$$v(x, y, 0) = -\cos(x)\sin(y) \tag{4.45}$$

$$p(x, y, 0) = -\frac{1}{4}(\cos(2x) + \cos(2y)) \tag{4.46}$$

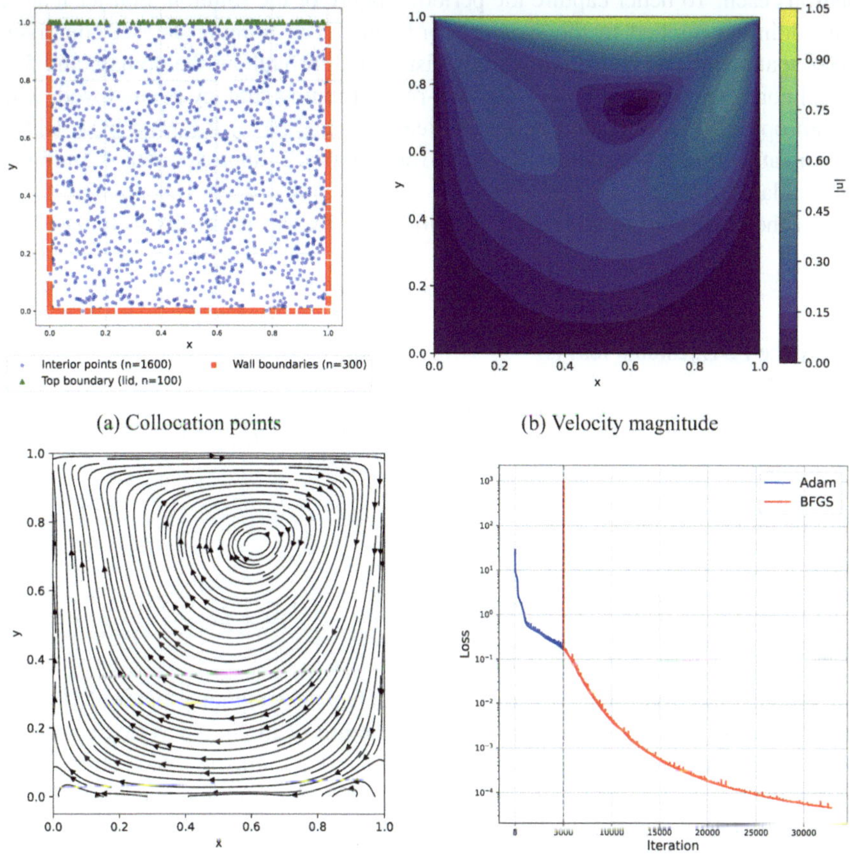

(a) Collocation points

(b) Velocity magnitude

(c) Streamlines

(d) Loss convergence

Fig. 4.14 PINN solution for lid-driven cavity flow at Re = 100. (**a**) Distribution of collocation points showing interior PDE points and boundary condition enforcement points. (**b**) Velocity magnitude contours revealing the flow intensity. (**c**) Streamlines illustrating the primary vortex structure centered at approximately (0.617, 0.734). (**d**) Loss convergence history showing the two-phase optimization strategy with Adam followed by BFGS

The exact solution is:

$$u(x, y, t) = \sin(x)\cos(y)e^{-2t/Re} \tag{4.47}$$

$$v(x, y, t) = -\cos(x)\sin(y)e^{-2t/Re} \tag{4.48}$$

$$p(x, y, t) = -\frac{1}{4}(\cos(2x) + \cos(2y))e^{-4t/Re} \tag{4.49}$$

For the time-dependent Taylor-Green vortex, the temporal domain extends from $t = 0$ to $t = 5$ at Reynolds number $Re = 100$. The neural network architecture consists of a single network with 5 hidden layers containing 64 neurons each. To better capture the periodic nature of the solution, Fourier feature embedding is applied to the spatial coordinates. The training dataset comprises 2000 spatiotemporal collocation points distributed throughout the domain and 500 initial condition points at $t = 0$. The two-phase training strategy is employed, with Adam optimization for 5000 iterations (learning rate 10^{-3}) followed by L-BFGS optimization for 5000 iterations. The total training time is approximately 12 minutes on standard hardware.

For the Taylor-Green vortex, the total loss function is specifically formulated as:

$$\mathcal{L}_{total} = \mathcal{L}_{PDE} + 10.0 \cdot \mathcal{L}_{IC} \tag{4.50}$$

where the PDE residual loss is:

$$\mathcal{L}_{PDE} = \frac{1}{N_{PDE}} \sum_{i=1}^{N_{PDE}} \left(|r_u^i|^2 + |r_v^i|^2 + |r_{cont}^i|^2 \right) \tag{4.51}$$

and the initial condition loss is:

$$\mathcal{L}_{IC} = \frac{1}{N_{IC}} \sum_{i=1}^{N_{IC}} \left(|u_\theta - u_{exact}|^2 + |v_\theta - v_{exact}|^2 + |p_\theta - p_{exact}|^2 \right)_{t=0} \tag{4.52}$$

Note that the Taylor-Green vortex uses an enhanced weighting factor of 10.0 for the initial condition loss to ensure accurate capture of the initial vortex structure, and no explicit boundary condition loss is needed due to the periodic nature of the problem which is naturally enforced through the Fourier feature embeddings.

Figures 4.15 through 4.17 illustrate the PINN performance for the Taylor-Green vortex problem. Figure 4.15 presents the velocity magnitude evolution at four time instances ($t = 0.0, 1.0, 2.5, 5.0$). The top row shows the PINN predictions, which capture the characteristic four-vortex pattern and its exponential decay due to viscous dissipation. The middle row displays the exact analytical solution for comparison, showing almost indistinguishable results to the predictions. The bottom row quantifies the absolute error, showing that errors remain below 10^{-2} throughout the simulation period, with the largest errors occurring near the vortex centers where velocity gradients are highest. The error magnitude increases slightly with time, which is expected as the network must capture increasingly diffused flow features.

Figure 4.16 focuses on the vorticity field evolution, a more sensitive measure of solution accuracy as it involves spatial derivatives of the velocity field. The PINN successfully captures the alternating positive and negative vorticity regions and their decay over time. The error patterns in the bottom row show that the PINN maintains accuracy even for this derived quantity, with errors concentrated at the vorticity extrema. The preservation of the vorticity structure demonstrates that the PINN not only learns the velocity field but also its derivatives accurately.

Figure 4.17 provides comprehensive quantitative analysis of the solution accuracy and training process. Panel (a) shows the training loss history, clearly illustrat-

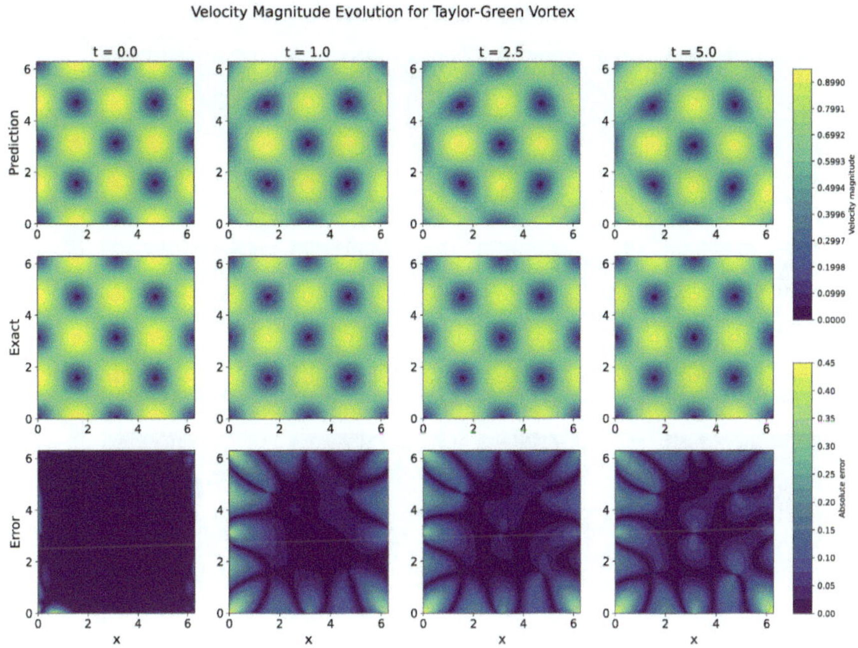

Fig. 4.15 Velocity magnitude comparison for the Taylor-Green vortex at Re = 100. Top row: PINN predictions. Middle row: Exact solution. Bottom row: Absolute error. The results are shown at four time instances: $t = 0.0, 1.0, 2.5, 5.0$, demonstrating the PINN's ability to accurately capture the flow evolution

ing the two-phase optimization strategy. The Adam optimizer provides rapid initial convergence over 5000 iterations, reducing the loss by several orders of magnitude. The subsequent L-BFGS phase, also running for 5000 iterations, refines the solution to achieve a final loss below 10^{-5}. Panel (b) tracks the temporal evolution of errors, showing both the L^2 relative error and kinetic energy error percentage. The L^2 error remains below 6×10^{-3} throughout the simulation, while the kinetic energy error stays below 0.5%, demonstrating excellent conservation properties of the PINN solution. Panel (c) visualizes the three-dimensional distribution of training points in the spatiotemporal domain. The blue PDE collocation points are uniformly distributed throughout the (x, y, t) space, ensuring adequate coverage of the entire solution domain. The red initial condition points are concentrated on the $t = 0$ plane, enforcing the exact initial vortex configuration. The wireframe grid at $t = 0$ provides a reference for the initial condition plane. This visualization demonstrates how PINNs handle time-dependent problems by treating time as an additional spatial dimension, allowing simultaneous solution across the entire temporal domain rather than traditional time-stepping approaches.

The results demonstrate that PINNs can effectively solve both steady-state and time-dependent incompressible Navier-Stokes equations. Key observations include:

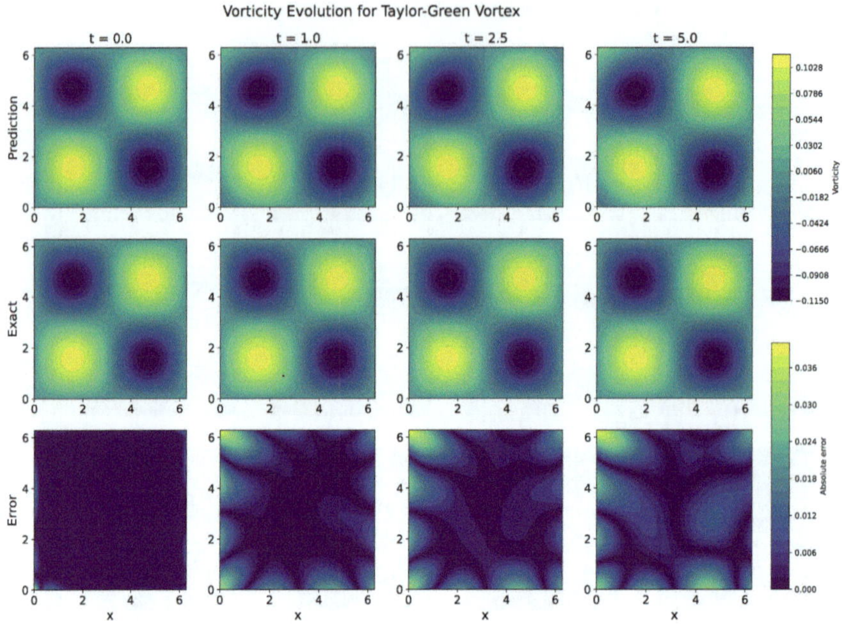

Fig. 4.16 Vorticity evolution for the Taylor-Green vortex at Re = 100. Top row: PINN predictions. Middle row: Exact solution. Bottom row: Absolute error. The PINN accurately captures the exponential decay of vorticity due to viscous dissipation at $t = 0.0, 1.0, 2.5, 5.0$

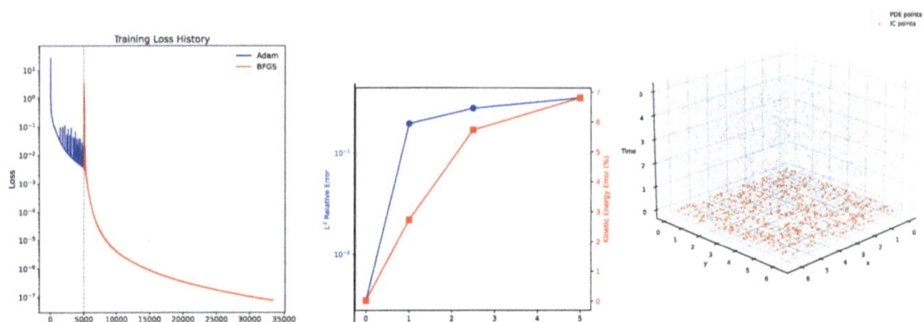

(a) Loss convergence history (b) Error evolution over time (c) 3D visualization of collocation points

Fig. 4.17 Training and error analysis for Taylor-Green vortex PINN. (**a**) Loss convergence history showing the two-phase optimization strategy with Adam followed by L-BFGS. (**b**) Evolution of L^2 relative error and kinetic energy error over time, showing excellent conservation properties. (**c**) 3D visualization of collocation points with PDE points (blue) distributed throughout the spatiotemporal domain $(x, y, t) \in [0, 2\pi] \times [0, 2\pi] \times [0, 5]$, and initial condition points (red) concentrated at $t = 0$. The wireframe grid indicates the initial condition plane

1. **Steady-state problems**: The lid-driven cavity results at Re = 100 show excellent agreement with benchmark data, with an average velocity error of only 1.26% compared to [9]. The PINN accurately captures the primary vortex structure and flow features.

2. **Time-dependent problems**: The Taylor-Green vortex solution maintains high accuracy over extended time periods, with relative errors below 1% for most of the simulation.
3. **Conservation properties**: PINNs naturally enforce mass conservation through the continuity equation loss, resulting in divergence-free velocity fields.
4. **Computational efficiency**: The cavity flow solution converged in approximately 8 minutes using the two-phase optimization strategy, and once trained, provides a continuous solution representation that can be evaluated at arbitrary points without interpolation.

Several challenges were addressed in implementing PINNs for the Navier-Stokes equations. The pressure-velocity coupling was handled through a single unified network with pressure regularization and enhanced boundary condition weighting to maintain numerical stability. The two-phase optimization strategy with Adam followed by L-BFGS optimization proved highly effective, achieving a final loss of 4.673×10^{-5} for the lid-driven cavity problem. Computational efficiency was achieved through optimized network architecture and training strategies, obtaining good accuracy (1.26% average error) in approximately 8 minutes. For the Taylor-Green vortex, periodic boundary conditions were handled through Fourier feature embeddings that encode the periodic structure directly into the network inputs. Continuity enforcement was enhanced by applying a weighted continuity residual with a 10× penalty factor, ensuring better mass conservation throughout the domain.

4.7 Applications to Hyperbolic PDEs

4.7.1 Wave Equation

The wave equation is a fundamental partial differential equation that describes the propagation of waves through various media, from sound waves in air to seismic waves in the Earth's crust. In this section, we demonstrate how Physics-Informed Neural Networks (PINNs) can be effectively applied to solve the wave equation, by incorporating the physical laws into the neural network training process. This approach offers an alternative to traditional numerical methods, particularly in scenarios where data-driven approaches can be combined with physics-based constraints.

The 2D wave equation in its general form is given by

$$\frac{\partial^2 u}{\partial t^2} = c^2 \left(\frac{\partial^2 u}{\partial x^2} + \frac{\partial^2 u}{\partial y^2} \right) \tag{4.53}$$

where $u(x, y, t)$ represents the displacement (or wave amplitude) at position (x, y) and time t, and c is the wave speed. For our benchmark problem, we consider the wave equation on a square domain $\Omega = [0, 1] \times [0, 1]$ with unit wave speed $c =$

1. The problem is subject to initial conditions $u(x, y, 0) = \sin(\pi x)\sin(\pi y)$ and $\frac{\partial u}{\partial t}(x, y, 0) = 0$, representing an initial displacement with zero initial velocity. The boundary conditions are homogeneous Dirichlet conditions, $u(x, y, t) = 0$ on $\partial\Omega$, meaning the wave amplitude is fixed at zero along all boundaries. This particular problem admits the analytical solution

$$u(x, y, t) = \sin(\pi x)\sin(\pi y)\cos\left(\sqrt{2}\pi t\right) \tag{4.54}$$

which describes a standing wave pattern that oscillates in time.

To solve the wave equation using PINNs, we approximate the solution $u(x, y, t)$ with a neural network $u_\theta(x, y, t)$, where θ represents the network parameters. The key insight of the PINN approach is to construct a loss function that enforces the PDE residual at collocation points inside the domain, the initial conditions, and the boundary conditions simultaneously. This is achieved through a composite loss function with four main components.

The PDE residual loss measures how well the neural network satisfies the wave equation at randomly sampled collocation points (x_i, y_i, t_i) inside the domain:

$$\mathcal{L}_{\text{PDE}} = \frac{1}{N_{\text{PDE}}}\sum_{i=1}^{N_{\text{PDE}}}\left|\frac{\partial^2 u_\theta}{\partial t^2}(x_i, y_i, t_i) - c^2\left(\frac{\partial^2 u_\theta}{\partial x^2}(x_i, y_i, t_i) + \frac{\partial^2 u_\theta}{\partial y^2}(x_i, y_i, t_i)\right)\right|^2 \tag{4.55}$$

The initial displacement loss enforces the prescribed initial shape:

$$\mathcal{L}_{\text{IC}_u} = \frac{1}{N_{\text{IC}}}\sum_{i=1}^{N_{\text{IC}}}|u_\theta(x_i, y_i, 0) - \sin(\pi x_i)\sin(\pi y_i)|^2 \tag{4.56}$$

The initial velocity loss ensures the wave starts from rest:

$$\mathcal{L}_{\text{IC}_v} = \frac{1}{N_{\text{IC}}}\sum_{i=1}^{N_{\text{IC}}}\left|\frac{\partial u_\theta}{\partial t}(x_i, y_i, 0)\right|^2 \tag{4.57}$$

The boundary condition loss enforces the Dirichlet conditions:

$$\mathcal{L}_{\text{BC}} = \frac{1}{N_{\text{BC}}}\sum_{i=1}^{N_{\text{BC}}}|u_\theta(x_i, y_i, t_i)|^2 \tag{4.58}$$

where (x_i, y_i, t_i) are points sampled on the boundary $\partial\Omega \times [0, T]$.

The total loss is a weighted sum of these components:

$$\mathcal{L}_{\text{total}} = w_{\text{PDE}}\mathcal{L}_{\text{PDE}} + w_{\text{IC}_u}\mathcal{L}_{\text{IC}_u} + w_{\text{IC}_v}\mathcal{L}_{\text{IC}_v} + w_{\text{BC}}\mathcal{L}_{\text{BC}} \tag{4.59}$$

Our implementation uses a fully connected neural network architecture with three hidden layers of 40 neurons each, employing hyperbolic tangent activation functions. The network takes three inputs (x, y, t) and outputs the displacement u. The boundary conditions are enforced through the network architecture by multiplying the output with $x(1-x)y(1-y)$, which automatically ensures zero values at the boundaries, and eliminates the need for the $\mathcal{L}_{BC}$ term. The required derivatives for the PDE residual are computed using automatic differentiation, which is efficiently implemented in JAX.

The training process employs a two-stage optimization strategy that has proven effective for PINN problems. First, we use the Adam optimizer for 15,000 iterations with a learning rate of 10^{-3} to quickly converge to a good solution region. Then, we switch to the L-BFGS optimizer from TensorFlow Probability for 3000 iterations of fine-tuning, as quasi-Newton methods often provide better convergence for the highly nonlinear optimization landscapes typical of PINN training. The collocation points consist of 8000 interior points, 1000 initial condition points, and 1000 boundary points, with loss weights set to $w_{\text{PDE}} = 1.0$, $w_{\text{IC}_u} = 10.0$, and $w_{\text{IC}_v} = 10.0$ to balance the different loss components.

Figure 4.18 shows the 3D distribution of collocation points used in training, with interior points shown in blue, initial condition points in red, and boundary points

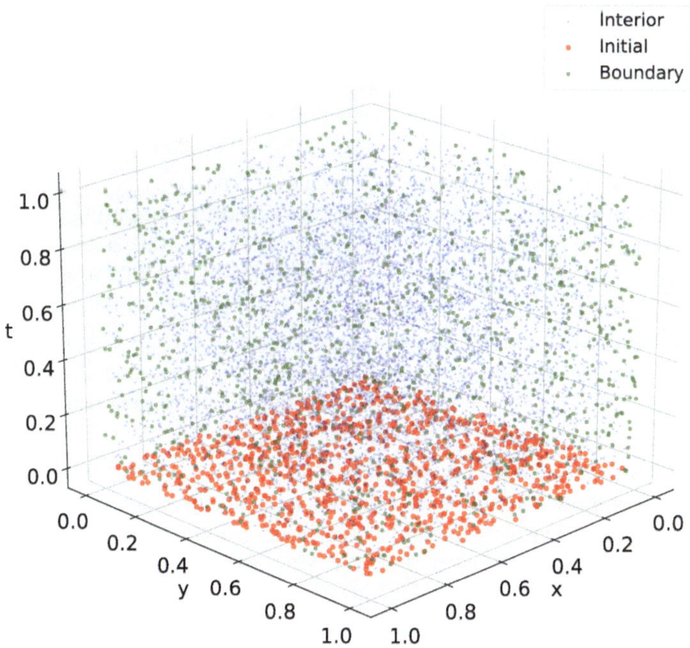

Fig. 4.18 3D scatter plot of collocation points used in PINN training. Blue points represent interior collocation points, red points show initial condition enforcement at $t = 0$, and green points indicate boundary condition enforcement locations

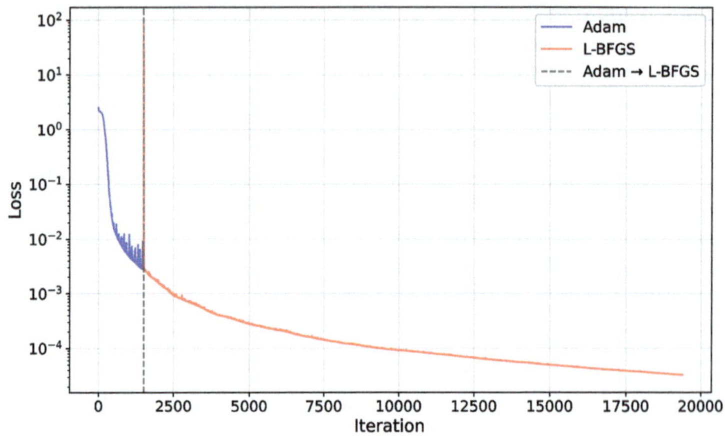

Fig. 4.19 Training loss history showing both Adam (blue) and L-BFGS (red) optimization phases. The vertical dashed line marks the transition between optimizers

in green. This visualization demonstrates how the points are distributed throughout the spatiotemporal domain to ensure adequate coverage for enforcing the physics constraints.

The evolution of the training loss during both optimization phases is shown in Figure 4.19. The transition between the Adam and L-BFGS optimizers is marked with a vertical dashed line, clearly showing the rapid initial convergence with Adam followed by the fine-tuning achieved with L-BFGS. The loss decreases by several orders of magnitude during training, indicating successful learning of the wave dynamics.

The performance of the trained PINN is evaluated by comparing its predictions with the exact solution at four time snapshots: $t = 0.0, 0.25, 0.5,$ and 1.0. Figure 4.20 presents a comprehensive comparison of the PINN solution. Panel (a) shows the PINN predictions across these time points, panel (b) displays the corresponding exact solutions, and panel (c) visualizes the absolute errors between predictions and exact solutions. The results demonstrate that the PINN successfully captures the wave propagation dynamics, with the standing wave pattern oscillating between positive and negative amplitudes as expected.

Quantitative error analysis reveals that the relative L2 error remains below 5% throughout the entire simulation period. At $t = 0$, the error is less than 0.1%, demonstrating excellent satisfaction of the initial conditions. For later times, the relative L2 error stays between 3% and 5%, with maximum pointwise errors remaining below 0.1. These error levels are quite acceptable for many practical applications and demonstrate the effectiveness of the PINN approach for wave propagation problems.

The total training time for this configuration is approximately 20–30 minutes on a Nvidia A4000 GPU, making the approach computationally feasible. Important factors to ensure that the network converges in a reasonable time are the two-

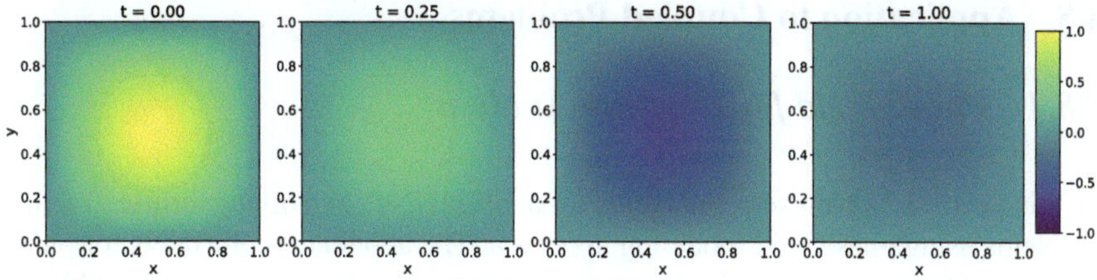

(a) PINN predictions

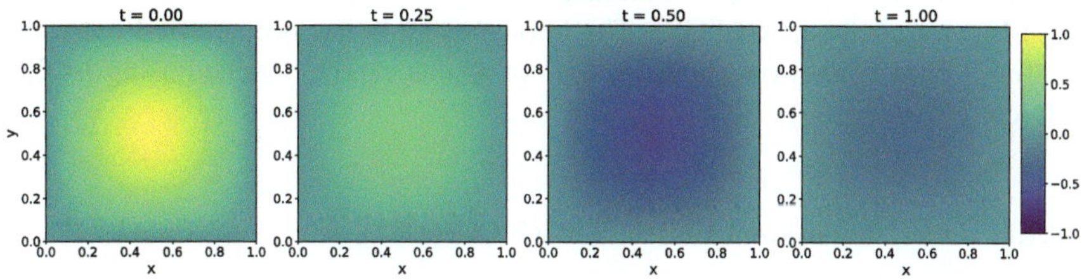

(b) Exact solutions

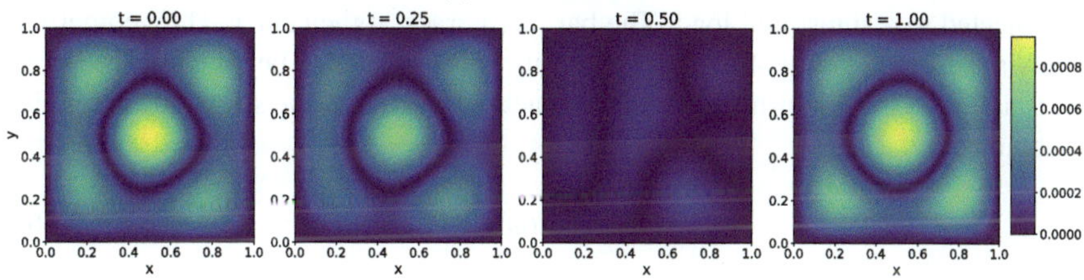

(c) Absolute errors

Fig. 4.20 Wave equation solution comparison at four time snapshots ($t = 0.0, 0.25, 0.5, 1.0$). (**a**) PINN predictions showing the wave displacement with color scale ranging from -1 to 1, capturing the oscillating standing wave pattern. (**b**) Exact analytical solutions of the wave equation for validation. (**c**) Absolute errors between PINN predictions and exact solutions, demonstrating that errors remain small throughout the simulation period. Each row uses a single colorbar for all time snapshots to facilitate comparison

stage optimization strategy, the balancing of loss weights, and the enforcement of boundary conditions through the network architecture itself.

This implementation demonstrates that PINNs can accurately solve the wave equation with relative errors below 5% while maintaining computational efficiency. The method captures the wave propagation dynamics well throughout the entire time domain and provides a mesh-free alternative to traditional numerical methods. The approach is particularly valuable in scenarios where combining data-driven methods with physics-based constraints is advantageous, such as when dealing with incomplete data or complex geometries. The complete JAX implementation is available in the accompanying Python file `wave_equation_pinn.py`, which includes the neural network definition with automatic boundary condition enforcement, loss function implementation for all components, the two-stage optimization strategy, and visualization utilities for results analysis.

4.8 Application to Coupled Problems

4.8.1 *Phase Field for Fracture Analysis*

In this section, we present an application of PINNs for solving brittle fracture problems using phase field modeling approach. The application of PINNs for phase field based fracture analysis is shown through three problems. For all the problems, PINNs is trained by using a combination of Adam optimizer and second-order quasi-Newton method (L-BFGS). The implementation has been carried out using the TensorFlow framework, however similar results can be obtained using JAX or Pytorch.

4.8.1.1 One-Dimensional Elastic Bar with Crack

In this example, a one-dimensional bar is considered that is fixed at both the ends and is subjected to a sinusoidal load. The bar has a crack at the center. The geometrical setup is presented in Fig. 4.21. For simplicity E is assumed to be unity and the strain (ϵ) is assumed to be non-negative in the crack zone. Hence, the stress-strain relation is obtained by:

$$\sigma = g(\phi)\epsilon. \tag{4.60}$$

$H_0(x)$ is the initial history field and is defined as:

$$H(x,0) = \begin{cases} 1000 & d(x,l) \leqslant l_0 \\ 0 & d(x,l) > l_0 \end{cases}, \tag{4.61}$$

where l_0 is considered 0.0125. In the fully cracked scenario, the analytical solutions for the displacement field represented as u_{ex} [10] and phase-field denoted by ϕ_{ex} [11] are represented as:

$$u_{ex} = \begin{cases} \frac{1}{\pi^2}\sin(\pi x) - \frac{(1+x)}{\pi} & \text{if } x < 0 \\ \frac{1}{\pi^2}\sin(\pi x) + \frac{(1+x)}{\pi} & \text{if } x > 0 \end{cases}. \tag{4.62a}$$

$$\phi_{ex} = \exp\left(\frac{-|x-a|}{l_0}\right), \tag{4.62b}$$

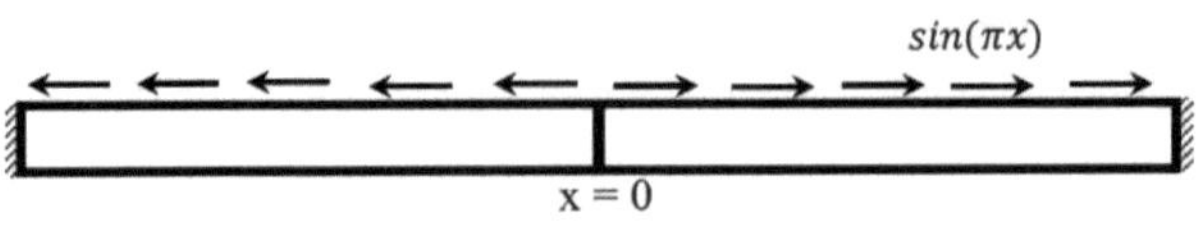

Fig. 4.21 Geometrical setup of one-dimensional elastic bar with crack

where the crack is located at a units. The Dirichlet boundary conditions are:

$$u(-1) = u(1) = 0, \tag{4.63}$$

where u is the solution of the elastic field in x-axis.

To solve this problem, a fully connected neural network with 3 hidden layers, comprising of 50 neurons in each hidden layer is considered. For the first two layers, the activation function is hyperbolic tangent, (`tanh`); whereas for the last layer, a linear activation function has been considered. The domain is subdivided into three sections along the x-axis; lc:$[-1.0, -2l_0]$, c:$[-2l_0, 2l_0]$, rc:$[2l_0, 1.0]$, where lc and rc represent the left and right side of the crack zone, respectively and c is the crack zone. This is done to generate more integration points in the vicinity of the crack. In each of the three sections, $N_x = 112$ Gauss points have been generated. In order to ensure that the neural network output exactly satisfies the Dirichlet boundary conditions, the outputs obtained from the network, $\hat{u}$ is modified as:

$$u = [(x + 1)(x - 1)]\hat{u}, \tag{4.64}$$

In the PINNs approach, the network is trained by minimizing the total variational energy of the system. In order to quantify the accuracy of the results, the relative $\mathcal{L}_2$ error and the root mean square error (RMSE) is obtained as:

$$\begin{aligned}\mathcal{L}_2^{rel} &= \frac{\sqrt{\sum_{i=1}^{N_{pred}} (f(x_i) - f_{ex}(x_i))^2 dx}}{\sqrt{\sum_{i=1}^{N_{pred}} (f_{ex}(x_i))^2 dx}}, \\ RMSE &= \sqrt{\frac{\sum_{i=1}^{N_{pred}} (f(x_i) - f_{ex}(x_i))^2}{N_{pred}}},\end{aligned} \tag{4.65}$$

where $f(x_i)$ denotes the value of the field variable obtained using the PINNs and $f_{ex}(x_i)$ is the exact value of the field variable. Figure 4.22a and b shows the displacement field, u and the phase-field, ϕ obtained using PINNs. Visually, the results obtained using PINNs overlaps with the analytical solutions obtained using Eqs. (4.62a) and (4.62b). To quantify the accuracy of the PINNs approach, the relative $\mathcal{L}_2$ error and RMSE corresponding to u and ϕ has been computed using (4.65). Corresponding to u and ϕ, a relative $\mathcal{L}_2$ prediction error of 4.46% and 3.61%, respectively have been observed while the observed RMSE error is 0.018 and 0.0544. As for the computational cost, PINNs require only 336 (3×112) Gauss points and 2100 (1500 + 600) iterations. The convergence history is shown in Fig. 4.22c.

Figure 4.23 shows the relative $\mathcal{L}_2$ prediction error for different number of integration points for both, u and ϕ. The relative $\mathcal{L}_2$ error decreases sharply with the increase in the number of integration points. This indicates the fast convergence

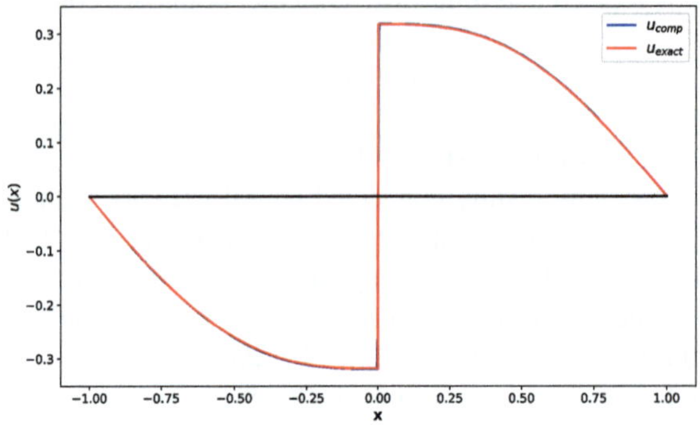

(a) Comparison of u_{exact} and u_{comp}.

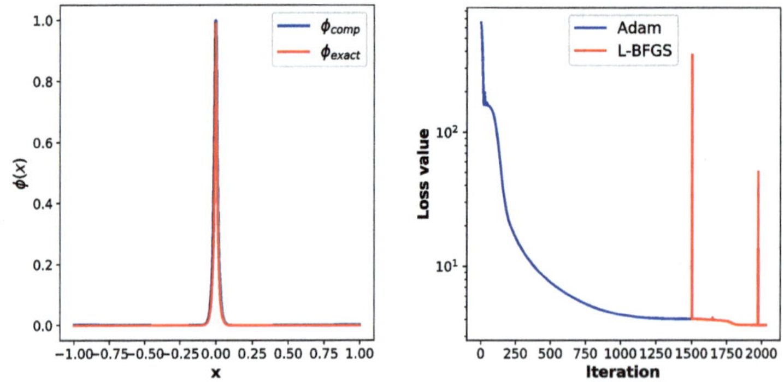

(b) Comparison of ϕ_{exact} and ϕ_{comp}. (c) Convergence of the loss function.

Fig. 4.22 Results for one-dimensional elastic bar with crack using variational energy based PINNs. (**a**) Comparison of u_{exact} and u_{comp}. (**b**) Comparison of ϕ_{exact} and ϕ_{comp}. (**c**) Convergence of the loss function

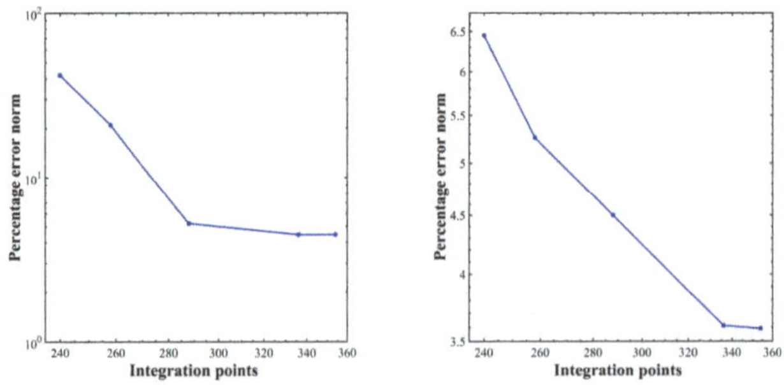

(a) Relative $\mathcal{L}_2$ norm for $\boldsymbol{u}$. (b) Relative $\mathcal{L}_2$ norm for ϕ.

Fig. 4.23 Change in the relative $\mathcal{L}_2$ error with the increase in the number of integration points for problem 1. The error sharply decreases with the increase in the number of integration points. (**a**) Relative $\mathcal{L}_2$ norm for $\boldsymbol{u}$. (**b**) Relative $\mathcal{L}_2$ norm for ϕ

rate of the proposed approach. Overall, the algorithm converges at 336 integration points.

4.8.1.2 Perforated and Notched Asymmetric Bending Example

In this example, the well known 'perforated and notched asymmetric three point bending' problem is considered. This problem has been previously analyzed both experimentally [12] and numerically [11]. It concerns an asymmetrically-notched beam with three holes. The geometrical setup and the boundary conditions for this example are depicted in Fig. 4.24a. This example has been typically chosen to show the performance of the variational energy based PINNs approach in predicting curved crack trajectories. Also, it illustrates the effectiveness of using NURBS for generating the geometry as the exact boundaries of the holes in the plate could be efficiently modeled. The domain is modeled with 29 NURBS patches which are subsequently refined along the path of expected crack propagation. The model generated using NURBS is shown in Fig. 4.24b. The material parameters are considered as $\lambda = 12.0$ kN/mm^2, $\mu = 8.0$ kN/mm^2, $G_c = 1 \times 10^{-3}$ kN/mm and $l_0 = 0.25$. Similar to [13], the left and right portion of the system are considered to be elastic.

We have considered a fully connected deep neural network with 3 hidden layers. Each hidden layer has 50 neurons. For the first two layers, `tanh` activation function and for the last layer, linear activation function has been considered. The plate has been discretized into 1184 elements with each element having 25 Gauss points. The crack is initiated using the strain history functional. The crack path is obtained by applying a constant displacement increment of $\Delta u = 1 \times 10^{-2}$. The Dirichlet boundary conditions for the problem are:

$$v(1,0) = v(19,0) = u(19,0) = 0, \quad v(10,8) = -\Delta v, \tag{4.66}$$

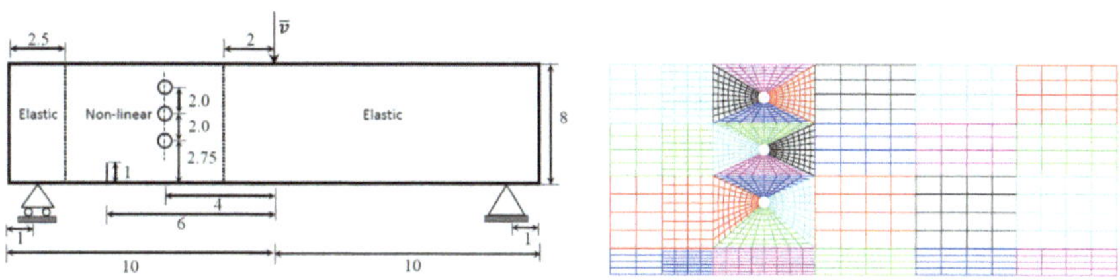

(a) Geometrical setup and boundary conditions. (b) Modeled geometry using NURBS.

Fig. 4.24 Perforated and notched asymmetric bending example. The diameter of the holes are 0.5. (**a**) Geometrical setup and boundary conditions. (**b**) Modeled geometry using NURBS where u and v are the solutions of the elastic field in the x and y-axis. To ensure that the neural network satisfies the boundary conditions exactly, the outputs of the elastic field are defined as:

$$\begin{aligned} u &= \frac{w_2}{(w_2+1)}\hat{u}, \\ v &= \frac{w_1 w_2 w_3}{(w_1+1)(w_2+1)(w_3+1)}\hat{v} + \frac{y}{8}\Delta v, \\ \text{where} \quad w_1 &= (x-1)^2 + y^2, \\ w_2 &= (x-19)^2 + y^2, \\ w_3 &= (x-10)^2 + (y-8)^2, \end{aligned} \tag{4.67}$$

where $\hat{u}$ and $\hat{v}$ are obtained from the neural network.

Figures 4.25 and 4.26 show the phase-field, ϕ and the predicted displacement field, v, respectively at certain displacement steps. For phase field, only the non-linear portion is shown. Similar to [13], the crack propagates through the second hole. However, propagation of the crack beyond the second hole has not been reported in the literature. We also note that while [13] obtained the results with 60,000 elements, the variational energy based PINNs require 1184 elements.

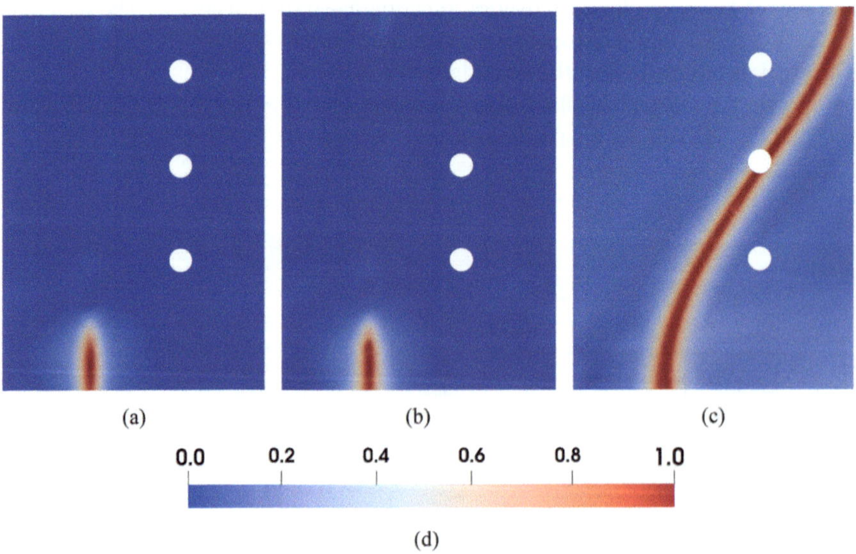

Fig. 4.25 Plots showing the predicted phase-field for prescribed displacement of (**a**) initialization of crack, (**b**) 1×10^{-2}, and (**c**) 16×10^{-2} using the proposed PINN approach, with (**d**) the corresponding horizontal color-bar

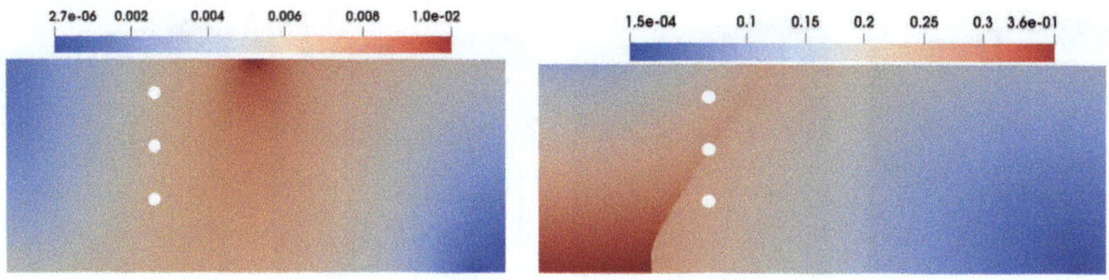

Fig. 4.26 Plots showing the predicted y-displacement for prescribed displacement of (**a**) 1×10^{-2}, and (**b**) 16×10^{-2} using the proposed PINN approach

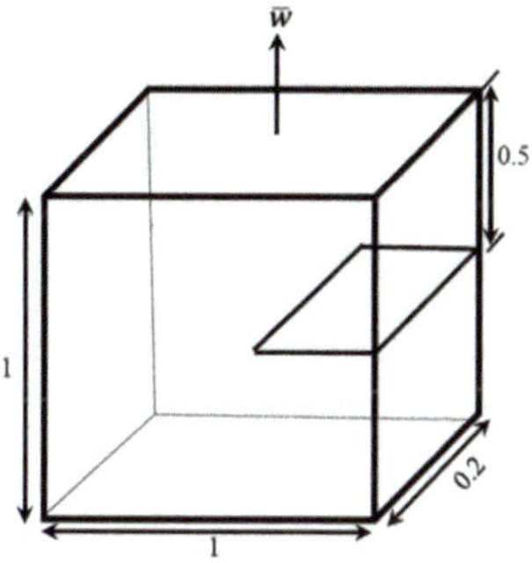

Fig. 4.27 Geometrical setup and boundary conditions of three dimensional mode-I tension test

4.8.1.3 Three Dimensional Mode-I Tension Test

In this example, a cube subjected to tensile loading is considered. The purpose of selecting this example is to illustrate the application of the variational energy based PINNs in predicting the crack path for three dimensional problems. The geometric setup and the associated boundary conditions are shown in Fig. 4.27. The NURBS mesh is used to generate the Gauss points. The material parameters considered are $\lambda = 12$ kN/mm^2, $\mu = 8$ kN/mm^2, $G_c = 0.5 \times 10^{-3}$ kN/mm and $l_0 = 0.0625$. To obtain the crack path, a constant displacement increment of $\Delta w = 1 \times 10^{-3}$ mm has been applied.

Similarly to the previous examples on this section, a fully connected deep neural network with 3 hidden layers is considered, comprising of 50 neurons in each hidden layer. For the first two layers, we have considered the `tanh` activation function; whereas for the last layer, linear activation function has been considered.The overall problem domain has been divided in to 512 elements with each element having 64 Gauss points. The crack is initiated using the strain history functional. The Dirichlet boundary conditions for the problem are:

$$u(x, y, 0) = v(x, y, 0) = w(x, y, 0) = 0, \quad w(x, y, 1) = \Delta w, \tag{4.68}$$

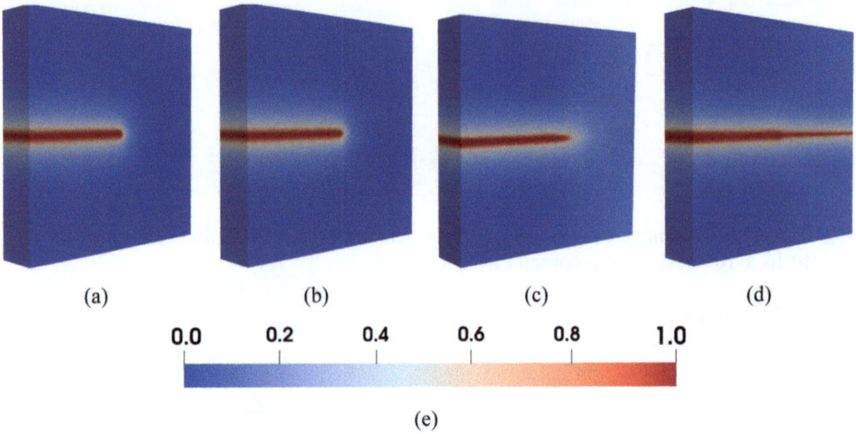

Fig. 4.28 Plots of cube showing the predicted phase-field for prescribed displacement of (**a**) initial crack (without any applied displacement), (**b**) $1 \times 10-3$, (**c**) $3 \times 10-3$, and (**d**) $6 \times 10-3$. Plot (**a**) shows the initialization of the crack. Plot (**e**) shows the color bar of the phase-field

where u, v and w are the solutions of the elastic field in x, y, and z-axis, respectively. To ensure that the boundary conditions are exactly satisfied, $\hat{u}$, $\hat{v}$ and $\hat{w}$ are modified as:

$$\begin{aligned} u &= z\hat{u}, \\ v &= z\hat{v}, \\ w &= z(z-1)\hat{w} + z\Delta u, \end{aligned} \tag{4.69}$$

The phase-field at certain displacement steps obtained using the variational energy based PINNs approach is shown in Fig. 4.28. The crack propagation pattern is found to be similar to those reported in literature [13, 14].

4.8.2 Allen-Cahn Equation

The Allen-Cahn equation is a reaction-diffusion equation that describes the process of phase separation in multi-component alloy systems and has applications in material science, image processing, and pattern formation. The equation models the evolution of a phase field variable $u(x, y, t)$ that represents the concentration or phase indicator function. For a two-dimensional domain $\Omega \subset \mathbb{R}^2$ with boundary $\partial\Omega$, the Allen-Cahn equation is given by:

$$\frac{\partial u}{\partial t} = \epsilon^2 \nabla^2 u - f(u) = \epsilon^2 \left(\frac{\partial^2 u}{\partial x^2} + \frac{\partial^2 u}{\partial y^2} \right) - f(u) \tag{4.70}$$

where $\epsilon > 0$ is a parameter related to the interface width between phases, and $f(u)$ is the derivative of a double-well potential. The most common choice is $f(u) = u^3 - u$, which corresponds to the double-well potential $F(u) = \frac{1}{4}(u^2 - 1)^2$ with minima at $u = \pm 1$ representing two stable phases. The equation is subject to initial and boundary conditions:

$$u(x, y, 0) = u_0(x, y), \quad \text{for } (x, y) \in \Omega \tag{4.71}$$

$$\frac{\partial u}{\partial n}(x, y, t) = 0, \quad \text{for } (x, y) \in \partial\Omega, \ t > 0 \tag{4.72}$$

where $u_0(x, y)$ is the initial phase field distribution and we impose homogeneous Neumann boundary conditions (no-flux condition) to model an isolated system.

The PINN approach for solving the Allen-Cahn equation employs a neural network $u_\theta(x, y, t)$ parameterized by weights θ to approximate the solution. The network is trained by minimizing a composite loss function that enforces the physics constraints:

$$\mathcal{L}_{total} = w_{PDE}\mathcal{L}_{PDE} + w_{IC}\mathcal{L}_{IC} + w_{BC}\mathcal{L}_{BC} \tag{4.73}$$

where w_{PDE}, w_{IC}, and w_{BC} are weighting factors that balance the contribution of each loss component. The PDE residual loss ensures that the neural network solution satisfies the Allen-Cahn equation at sampled collocation points within the spatiotemporal domain:

$$\mathcal{L}_{\text{PDE}} = \frac{1}{N_{\text{PDE}}} \sum_{i=1}^{N_{\text{PDE}}} \Bigg| \frac{\partial u_\theta}{\partial t}(x_i, y_i, t_i) - \epsilon^2 \Big(\frac{\partial^2 u_\theta}{\partial x^2}(x_i, y_i, t_i) + \frac{\partial^2 u_\theta}{\partial y^2}(x_i, y_i, t_i) \Big) + u_\theta^3(x_i, y_i, t_i) - u_\theta(x_i, y_i, t_i) \Bigg|^2. \tag{4.74}$$

where (x_i, y_i, t_i) are randomly sampled collocation points in $\Omega \times [0, T]$. The initial condition loss enforces the prescribed initial phase field distribution:

$$\mathcal{L}_{IC} = \frac{1}{N_{IC}} \sum_{i=1}^{N_{IC}} |u_\theta(x_i, y_i, 0) - u_0(x_i, y_i)|^2 \tag{4.75}$$

where (x_i, y_i) are points sampled in the spatial domain Ω at $t = 0$. For Neumann boundary conditions, the boundary loss enforces zero normal derivative at the boundaries:

$$\mathcal{L}_{BC} = \frac{1}{N_{BC}} \sum_{i=1}^{N_{BC}} \left| \frac{\partial u_\theta}{\partial n}(x_i, y_i, t_i) \right|^2 \tag{4.76}$$

where (x_i, y_i, t_i) are points sampled on the boundary $\partial\Omega \times [0, T]$ and $\frac{\partial}{\partial n}$ denotes the normal derivative. For rectangular domains, this simplifies to enforcing $\frac{\partial u_\theta}{\partial x} = 0$ on vertical boundaries and $\frac{\partial u_\theta}{\partial y} = 0$ on horizontal boundaries.

To demonstrate the effectiveness of PINNs for solving the Allen-Cahn equation, we consider a benchmark problem on the unit square $\Omega = [-1, 1] \times [-1, 1]$ with $\epsilon = 0.01$ and $T = 10$. This choice of ϵ creates sharp interfaces between phases while remaining computationally tractable. We initialize the phase field with a circular region of the positive phase:

$$u_0(x, y) = \tanh\left(\frac{r_0 - \sqrt{x^2 + y^2}}{\sqrt{2}\epsilon}\right) \tag{4.77}$$

where $r_0 = 0.5$ is the initial radius of the circular interface. This initial condition represents a smooth transition between $u = 1$ inside the circle and $u = -1$ outside, with the transition width controlled by ϵ. The hyperbolic tangent profile is the exact stationary solution for a planar interface in the Allen-Cahn equation.

The Allen-Cahn dynamics cause the circular interface to shrink over time due to surface tension effects. The interface velocity is proportional to the mean curvature, leading to the well-known mean curvature flow in the sharp interface limit ($\epsilon \to 0$). For our circular geometry, the radius decreases according to $\frac{dr}{dt} \approx -\frac{\epsilon}{r}$, which provides a semi-analytical reference for validating the PINN solution.

The PINN implementation employs a fully connected neural network with an input layer of 3 neurons (x, y, t), four hidden layers with 50 neurons each using tanh activation functions ($\sigma(x) = \tanh(x)$), and an output layer with 1 neuron representing the phase field u. The network is trained using a two-phase optimization strategy that begins with the Adam optimizer for 1000 iterations with a learning rate of 10^{-3} for rapid initial convergence, followed by L-BFGS optimization using TensorFlow Probability's implementation for 500 iterations to fine-tune the solution.

We sample 2000 collocation points for the PDE residual, 400 points for the initial condition, and 400 points for boundary conditions. The loss weights are set to $w_{PDE} = 1$, $w_{IC} = 100$, and $w_{BC} = 10$ to ensure proper enforcement of initial and boundary conditions. The implementation uses JAX with 64-bit precision to handle the multiple scales present in the Allen-Cahn equation. Figure 4.29 shows the training loss evolution during both optimization phases. The Adam optimizer provides rapid initial convergence, reducing the loss by three orders of magnitude in the first 1000 iterations. The subsequent L-BFGS optimization further refines the solution, achieving a final total loss of approximately 0.12.

Figure 4.30 presents the PINN solution at different time snapshots, showing the evolution of the phase field. The circular interface shrinks over time as expected from the mean curvature flow dynamics. The solution maintains sharp interfaces despite using a continuous neural network representation, demonstrating the ability of PINNs to capture steep gradients. The results demonstrate that PINNs can successfully solve the Allen-Cahn equation, capturing both the sharp interface dynamics and the correct shrinking behavior of the circular region. The

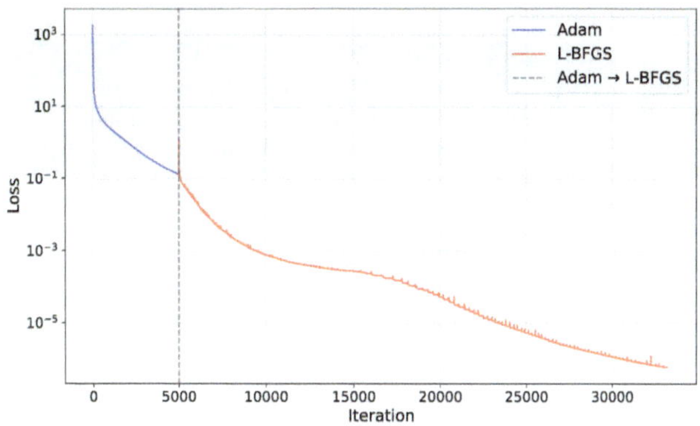

Fig. 4.29 Training loss history for the Allen-Cahn PINN showing the transition from Adam to L-BFGS optimization. The vertical dashed line at iteration 1000 marks the switch between optimizers. The loss is plotted on a logarithmic scale to show the full range of convergence

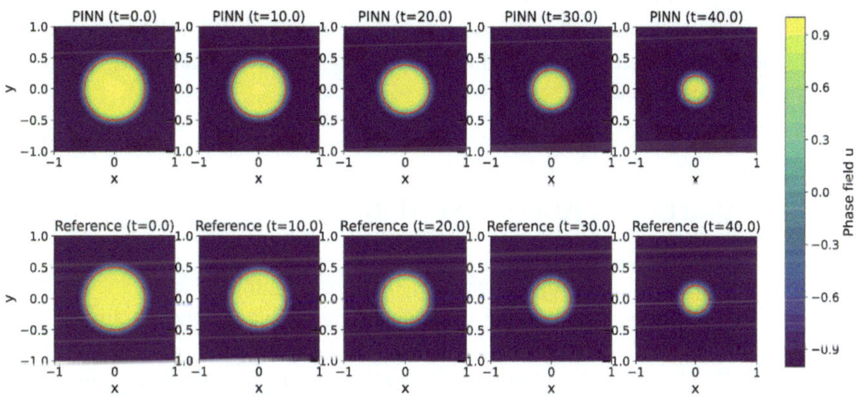

Fig. 4.30 Evolution of the phase field in the Allen-Cahn equation. Top row: PINN predictions at $t = 0, 2.5, 5, 7.5, 10$. Bottom row: Reference solution obtained using a high-resolution finite difference scheme. The colorbar indicates phase values ranging from -1 (blue) to $+1$ (red)

method achieves reasonable accuracy without requiring explicit interface tracking or adaptive mesh refinement, which are typically necessary for traditional numerical methods when ϵ is small. The errors are higher at later times as the interface becomes increasingly sharp, but the method still captures the qualitative behavior well (Fig. 4.31).

The accuracy of the PINN approach for the Allen-Cahn equation highlights a few advantages over traditional numerical methods. The mesh-free computation eliminates the need for structured grids or complex meshing procedures. Automatic differentiation through the neural network provides exact derivatives which is convenient in the computer implementation. The continuous and differentiable solution representation everywhere in the domain allows for post-processing and analysis at arbitrary points without interpolation. Furthermore, the method demonstrates stable

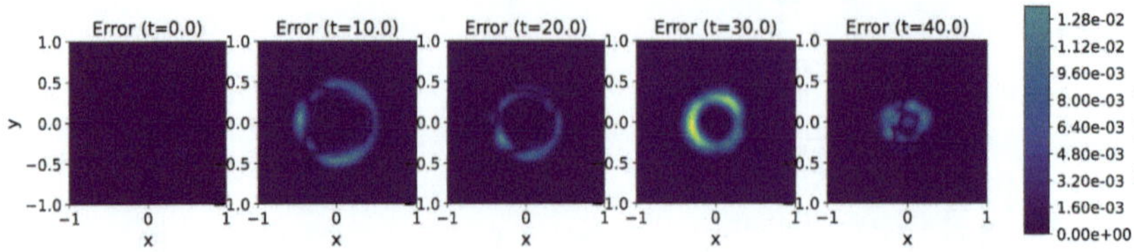

Fig. 4.31 Absolute error between PINN predictions and reference solution at different time instances. The maximum error remains below 0.05 throughout the simulation, with highest errors occurring near the sharp interface region

long-time integration without accumulation of discretization errors that often plague time-stepping schemes.

These results establish PINNs as a possible alternative to traditional numerical methods for phase field models, particularly for problems with complex geometries or when a smooth, differentiable solution representation is desired. While the accuracy could be improved with more training iterations and collocation points, the method demonstrates the feasibility of applying PINNs to challenging nonlinear PDEs with sharp gradients.

4.9 Comparison to Alternative Methods

As already pointed out before, there are numerous computational methods for the solutions of PDEs, which solve the underlying problem either in strong form (such as the FDM) or in weak form such as the FEM, certain meshfree methods and most IGA (Isogeometric Analysis) formulations. Due to their robustness and ability to model complex geometries, weak-form based methods are more popular in computational solid mechanics. They are the focus of this section.

Computational methods such as FEM transform the solution of a BVP to the solution of a linear (or linearized) system of equations (the latter one is commonly solved iteratively) while machine learning based solutions to PDEs solve a non-convex optimization problem. Even when the underlying problem is linear and well-posed—such as the BVP of linear elasticity, the ML based solution yields a nonlinear nonconvex optimization problem. Indeed, solving a non-convex optimization problem seems certainly more cumbersome than solving a linear (or linearized) system of equations. Furthermore, while there are convergence proofs for computational methods such as FEM, which guarantee a certified solution, i.e. the solution gets more accurate once the discretization is refined, such proofs are absent in the case of ML based solutions of PDEs. In fact, there are many parameters, which determine the accuracy and robustness of a ML based solution to PDEs, and they are difficult to calibrate. So, what is the advantage of a machine learning based solution of PDEs? First of all, not all problems in engineering are linear and well-posed. In fact, the most challenging problems are non-linear. Even when the underlying PDE is linear, material and/or geometrical nonlinearities in pure mechanical problems are quite standard. Secondly, there are many ill-posed problems in engineering; optimization and inverse analysis are classical examples. So, ML might not be a good alternative for all type of problems which are governed by PDEs but at least for some, i.e. for problems methods such as the FEM experience difficulties. These might include nonlinear PDEs, ill-posed problems, stochastic PDEs or problems with high dimensionality, where the finite element method scales poorly. Indeed, ML approaches build a natural framework for uncertainty analysis, optimization and inverse analysis. Nonetheless, according to the 'principle of modeling' (i.e. starting with simple problems first), it might be worth to 'start' with linear problems, where it is easier to gain a fundamental understanding.

4.9.1 Finite Element Method

The finite element method is among the most popular discretization methods to solve PDEs. The key idea is to subdivide the domain into subdomains, the elements. The primary field (often called trial function) and the associated test function is then interpolated with finite element shape functions. For instance, for the boundary value problem (BVP) of linear elastostatics (see the introduction in Chap. 3), the discretization of the displacement field reads:

$$\mathbf{u}^h(\mathbf{X}) = \sum_{I=1}^{n} N_I(\mathbf{X})\, \mathbf{u}_I = \mathbf{N}(\mathbf{X})\mathbf{d} \tag{4.78}$$

where $N_I(\mathbf{X})$ denotes the finite element shape function at node I, $\mathbf{u}_I$ is the nodal parameter at that node and n the number of nodes per element. Most FE shape functions are based on Lagrange polynomials, which are interpolatory, which means that the nodal parameter at the node is identical to the physical value at that node. Or in other words, $\mathbf{u}_I$ is the displacement at the node I for above mentioned BVP. For implementation purposes, the shape functions are stored in a matrix $\mathbf{N}(\mathbf{X})$ and the nodal parameters in the vector $\mathbf{d}$. As it is difficult to define shape functions for generic geometries in the physical space, they are commonly defined in a parameter space leading to the concept of parametric finite elements. If the geometry is discretized with the same shape functions as the trial functions, it is called an iso-parametric finite element formulation. Derivatives of the trial functions are then obtained simply by taking derivatives of the shape functions. For the BVP of linear elastostatics for instance, we need the linear strain tensor, which is the symmetric part of the displacement gradient. In vector matrix form, it can be written as

$$\boldsymbol{\epsilon}^h(\mathbf{X}) = \mathbf{B}(\mathbf{X})\mathbf{d} \tag{4.79}$$

where $\mathbf{B}$ is the matrix, which contains the derivatives of the shape functions. The same shape functions are often used to discretize the test functions—often denoted by δ yielding

$$\delta \mathbf{u}^h(\mathbf{X}) = \mathbf{N}(\mathbf{X})\delta \mathbf{d} \tag{4.80}$$

and for the virtual strains

$$\delta \boldsymbol{\epsilon}^h(\mathbf{X}) = \mathbf{B}(\mathbf{X})\delta \mathbf{d} \tag{4.81}$$

Rewriting the weak form, Eq. (3.14), in Voigt notation

$$\int_{\Gamma_t} \delta \mathbf{u}\, \bar{\mathbf{t}}\, d\Gamma + \int_{\Omega} \delta \mathbf{u}\, \mathbf{b}\, d\Omega - \int_{\Omega} \delta \boldsymbol{\epsilon}\, \boldsymbol{\sigma}\, d\Omega \tag{4.82}$$

and substituting the discretization of the test and trial functions (and their spatial derivatives) into Eq. (4.82) leads to

$$\sum_{e=1}^{M} \left(\int_{\Gamma_t^e} \left(\mathbf{N}^e \delta \mathbf{d}^e\right)^T \, d\Gamma + \int_{\Omega^e} \left(\mathbf{N}^e \delta \mathbf{d}^e\right)^T \mathbf{b} + \int_{\Omega^e} \left(\mathbf{B}^e \delta \mathbf{d}^e\right)^T \mathbf{C}\mathbf{N}^e \mathbf{d}^e \right) \tag{4.83}$$

After some algebra and assembly operations, the well known linear system of equations is obtained:

$$\mathbf{K}\,\mathbf{D} = \mathbf{F}_{ext} \tag{4.84}$$

where $\mathbf{K} = \sum_{e=1}^{M} \mathbf{k}^e$, $\mathbf{D} = \sum_{e=1}^{M} \mathbf{d}^e$ and $\mathbf{F}_{ext} = \sum_{e=1}^{M} \mathbf{f}^e_{ext}$ indicate the global stiffness matrix, the global vector of unknown and the global external force vector, respectively; M is the number of elements in the entire discretization. The local element stiffness matrix and the local element force vector is given by

$$\begin{aligned} \mathbf{k}^e &= \int_{\Omega^e} \mathbf{B}^T \mathbf{C} \mathbf{B}\, d\Omega \\ \mathbf{f}^e_{ext} &= \int_{\Omega^e} \mathbf{N}^T \mathbf{b}\, d\Omega + \int_{\Gamma_t^e} \mathbf{N}^T \bar{\mathbf{t}}\, d\Gamma \end{aligned} \tag{4.85}$$

where the components of the fourth order elasticity tensor are stored in the matrix $\mathbf{C}$. Note that for an iso-parametric FE formulation, the integrals are not evaluated in the physical domain but in the parameter space. Commonly, the integration is performed numerically by Gauss quadrature, which can exactly integrate a polynomial of arbitrary order. However, as the finite element shape functions are formulated in the parameter space, the computation of the matrix $\mathbf{B}$ requires the inverse of the Jacobi matrix, i.e. the mapping between the parameter and physical space. Hence, for generic cases, a rational function is actually integrated, which leads to integration errors. However, experience showed that these errors are quite small.

As mentioned before, the most typical shape functions are Lagrange polynomials ($l_i^{n-1}(x) = \prod_{j=1,j\neq i}^{n} \frac{x-\bar{x}_j}{\bar{x}_i-\bar{x}_j}$ in one dimensions, where the subscript refers to the node number and the superscript refers to the polynomial order with n being the number of nodes; the superimposed bar indicates the position of the specific node), which are defined in a parameter space, commonly chosen between 0 and 1 or -1 and 1. For the latter case, the shape function of a linear Lagrange element in 1D reads:

$$N_I(\xi) = \frac{1}{2}\,(1 \pm \xi) \tag{4.86}$$

where ξ is the local coordinate in the parameter space. One can clearly see that this shape functions fulfill the Kronecker delta property, i.e. $N_I(\xi_J) = \delta_{IJ}$. Hence, the nodal parameter is equivalent to the physical value at the node. It also facilitates the imposition of essential boundary conditions, which are commonly imposed in strong form in finite elements. The extension to higher order shape functions and higher dimensions is straight forward. For instance, two- and three-dimensional elements are obtain through a simple tensor product, such that the shape functions of a 4-node bi-linear quadrilateral (Q4) element are given by

$$N_I(\xi, \eta) = \frac{1}{4}\,(1 \pm \xi)\,(1 \pm \eta) \tag{4.87}$$

η being the coordinate perpendicular to ξ.

4.9.2 Mesh-Free Method

The key difference to the FE discretization is the way the shape functions are constructed. Most meshfree shape functions are rational which are based on a so-called weighting function. Figure 4.32 illustrates a common weighting function, the cubic spline, which can be computed by

$$\mathrm{W(r)} = \begin{cases} \frac{C}{h^D}\left(1 - 1.5z^2 + 0.75z^3\right) & 0 \leq z < 1 \\ \frac{C}{4\,h^D}\,(2 - z)^3 & 1 \leq z \leq 2 \\ 0 & z > 2 \end{cases} \tag{4.88}$$

where D is the dimension, $z = r/h$ and C is a constant depending on the dimension

$$\mathrm{C} = \begin{cases} 2/3 & D = 1 \\ 10/(7\,\pi) & D = 2 \\ 1/\pi & D = 3 \end{cases} \tag{4.89}$$

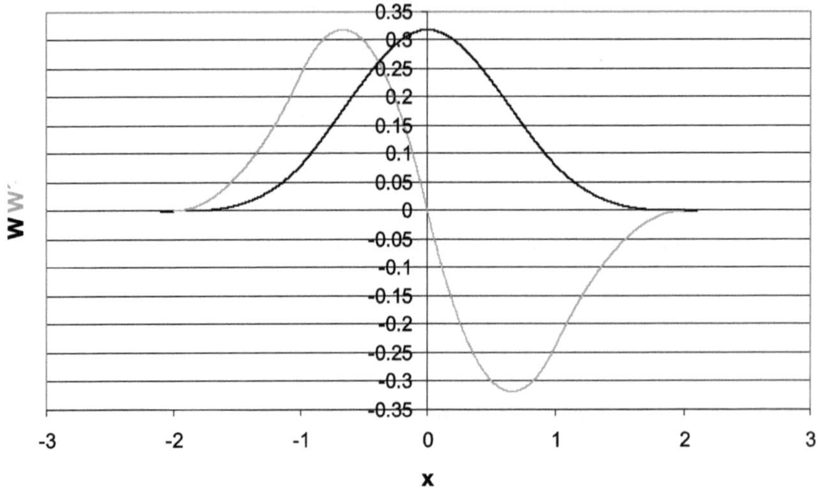

Fig. 4.32 Cubic spline in 1D and its first spatial derivative

In higher order dimensions and for a circular support, $r = \|\mathbf{x} - \mathbf{x}_J\|$. The kernel functions need to fulfill the following three requirements:

$$\lim_{h_0 \to 0} W(\mathbf{x}_I - \mathbf{x}_J, h_0) = \delta(\mathbf{x}_I - \mathbf{x}_J) \tag{4.90}$$

$$\int_{\Omega_0} W(\mathbf{x}_I - \mathbf{x}_J, h_0) d\Omega_0 = 1 \tag{4.91}$$

$$W(\mathbf{x}_I - \mathbf{x}_J, h_0) = 0 \qquad \|\mathbf{x}_I - \mathbf{x}_J\| \geq R \tag{4.92}$$

δ referring to the Dirac delta function and R is a shape parameter often related to h. The last condition is called compact support of the shape function and ensures computational efficiency though there are a few meshfree methods which employ a global support. Or in other words: R determines the support size, often called domain of influence, i.e. the domain a central nodes 'interacts' with its neighbors. It is an 'additional' (compared to FEM) parameter, which can vary from node to node. For specific interpolation/approximation problems and nodal arrangements, the optimal value for R can be computed. However, for the solution of PDEs, it is difficult to impossible to determine the optimal value and one relies on 'experience'. It has been shown, e.g. by [15, 16] that a good ratio between R and the nodal distance lies in the range of 2.8–3.2. Commonly, additional conditions are imposed on the kernel function, i.e.

$$W(\mathbf{x}_I - \mathbf{x}_J, h_0) = W(\mathbf{x}_J - \mathbf{x}_I, h_0) \tag{4.93}$$

$$\nabla_0 W(\mathbf{x}_I - \mathbf{x}_J, h_0) = -\nabla_0 W(\mathbf{x}_J - \mathbf{x}_I, h_0) \tag{4.94}$$

While the finite element method has a fixed connectivity between nodes and elements, the interaction between meshfree nodes can vary during the course of the simulation, which is advantageous for problems with large deformations, fracture and fragmentation.

The discretization of the test and trial functions in a meshfree method can be written as in FEM:

$$\mathbf{u}^h(\mathbf{x}) = \sum_{I=1}^{n} N_I(\mathbf{x})\,\mathbf{u}_I = \mathbf{N}(\mathbf{x})\mathbf{d} \tag{4.95}$$

where n denotes the number of neighbors in the domain of influence of a meshfree node. While for the same element type, the number of nodes per element does not change, they vary in a meshfree method. For instance, for an equispaced structured discretization with constant (identical) support sizes, nodes at the boundary have less neighbours than nodes in the interior. As already mentioned before, the 'key' difference to FEM lies in the computation of the shape functions $N_I(\mathbf{x})$. There are countless meshfree methods. Among the most popular meshfree methods based on the weak form are the element-free Galerkin (EFG) method [17] and Reproducing Kernel Particle Method (RKPM) [18], which show also many similarities, see e.g. [19, 20]. Here, let us focus on the EFG method, which is based on a Moving Least Squares (MLS) approximation, which can be expressed by

$$u_h(\mathbf{x}) = \sum_{J}^{n} \mathbf{p}(\mathbf{x}_J)\,\mathbf{a}(\mathbf{x}) \tag{4.96}$$

where $\mathbf{p}$ denotes a polynomial basis. The minimum requirement for a mechanical problem is that one can reproduce a rigid body motion and a constant strain state exactly. This requires a linear approximation, i.e.

$$\mathbf{p}(\mathbf{x}) = \begin{pmatrix} 1 & x & y & z \end{pmatrix} \quad \forall \mathbf{x} \in \mathrm{Re}^3 \tag{4.97}$$

The vector $\mathbf{a}$ depends on $\mathbf{x}$ (explaining the name 'moving' least squares) and is obtained through minimizing a discrete weighted L_2 error norm

$$\mathcal{L} = \sum_{J}^{n} \left(\mathbf{p}(\mathbf{x}_J)^T\,\mathbf{a}(\mathbf{x}) - u_J\right)^2 W(\mathbf{x} - \mathbf{x}_J, h) \tag{4.98}$$

Minimizing Eq. (4.98) with respect to $\mathbf{a}$ leads to

$$\mathbf{a}(\mathbf{x}) = \mathbf{A}(\mathbf{x})\mathbf{p}(\mathbf{x}_I)W(\mathbf{x} - \mathbf{x}_J, h_0) \tag{4.99}$$

with

$$\mathbf{A}(\mathbf{x}) = \sum_{I}^{n} \mathbf{p}(\mathbf{x}_J)\, \mathbf{p}^T(\mathbf{x}_J)\, W(\mathbf{x} - \mathbf{x}_J, h) \tag{4.100}$$

Substituting **a** back into the MLS approximation, Eq. (4.96), results in the final approximation

$$\mathbf{u}^h(\mathbf{x}) = \sum_{I=1}^{n} N_I(\mathbf{x})\, \mathbf{u}_I = \mathbf{N}(\mathbf{x})\mathbf{d} \tag{4.101}$$

with

$$N_J(\mathbf{x}) = \mathbf{p}(\mathbf{x}_J)^T \cdot \mathbf{A}(\mathbf{x})^{-1} \cdot \mathbf{p}(\mathbf{x}_J)\, W(\mathbf{x} - \mathbf{x}_J, h) \tag{4.102}$$

One can easily see that the computation of the MLS shape functions is computationally more demanding compared to the determination of the shape functions of the Lagrange elements. Moreover, meshfree methods such as the EFG method are not interpolatory, i.e. they do not fulfill the Kronecker delta property ($N_I(\mathbf{x}_J) \neq \delta_{IJ}$). This complicates the imposition of Dirichlet boundary conditions. The nodal parameters are also not any more the physical quantities at the node. For instance, to determine the displacement field at a meshfree node for an elasticity problem, one needs to employ Eq. (4.101). Meshfree methods based on the weak form require the evaluation of the integrals. For problems with small deformations, it is common to construct a background mesh (purely for the purpose of integration) and employ Gauss quadrature. Methods which do not require any background mesh have been coined as 'truly' meshfree, though it is not difficult to construct the background mesh for nearly any problem. Nonetheless, the rational character of the meshfree shape function increases integration errors—when Gauss quadrature is used. A brute force approach is to increase the number of quadrature points. An advantage of meshfree methods over finite elements are their higher continuity, that can be exploited for higher-order PDEs, see Chap. 3. Also problems with large deformations, fracture and fragmentation can be handled better by meshfree methods.

Applying the EFG method to the solution of the linear boundary value problem of elastostatics (in weak form) leads to Eq. (4.84) with different matrices **N** and **B** that contain now the meshfree shape functions and their spatial derivatives, respectively. Otherwise, there are subtle differences in the implementation. For instance, a meshfree method requires a neighbour search in order to find the neighbouring nodes (or Gauss points) for each 'central' node. If Gauss quadrature is used to evaluate the integrals, the position of the Gauss points in the physical domain is required, which can simply be obtained through the meshfree approximation. And while one commonly goes through the element list in order to assemble e.g. the global stiffness matrix in finite elements, there would be a loop over the nodes (and an inner loop over the neighbouring Gauss points) in meshfree methods.

4.9.3 Isogeometric Analysis

Isogeometric Analysis (IGA), introduced by Hughes et al. [21], aims to bridge the gap between Computer-Aided Design (CAD) and Finite Element Analysis (FEA) by employing the same basis functions used in CAD geometry representation directly for analysis. The fundamental idea is to use Non-Uniform Rational B-Splines (NURBS), which are the industry standard in CAD, as both the geometric representation and the basis for approximating the solution fields. This eliminates the need for mesh generation and the associated geometric approximation errors inherent in traditional FEM.

The key building blocks of IGA are B-spline basis functions, which are defined recursively using the Cox-de Boor formula. For a given knot vector $\Xi = \{\xi_1, \xi_2, \ldots, \xi_{n+p+1}\}$, where n is the number of basis functions and p is the polynomial degree, the B-spline basis functions are defined as:

$$N_{i,0}(\xi) = \begin{cases} 1 \text{ if } \xi_i \leq \xi < \xi_{i+1} \\ 0 \quad \text{otherwise} \end{cases} \tag{4.103}$$

$$N_{i,p}(\xi) = \frac{\xi - \xi_i}{\xi_{i+p} - \xi_i} N_{i,p-1}(\xi) + \frac{\xi_{i+p+1} - \xi}{\xi_{i+p+1} - \xi_{i+1}} N_{i+1,p-1}(\xi) \tag{4.104}$$

For multivariate problems, tensor products of univariate B-spline basis functions are employed. In 2D, the basis functions become:

$$R_{i,j}^{p,q}(\xi, \eta) = N_{i,p}(\xi) M_{j,q}(\eta) \tag{4.105}$$

where $M_{j,q}$ are B-spline basis functions in the η direction with polynomial degree q.

NURBS basis functions are obtained as the weighted projections of B-spline basis functions:

$$R_I(\xi) = \frac{N_{I,p}(\xi) w_I}{\sum_{J=1}^{n} N_{J,p}(\xi) w_J} \tag{4.106}$$

where w_I are the weights associated with control points. The geometry and the displacement field are then represented using the same NURBS basis:

$$\mathbf{X}(\xi) = \sum_{I=1}^{n} R_I(\xi) \mathbf{P}_I \tag{4.107}$$

$$\mathbf{u}^h(\xi) = \sum_{I=1}^{n} R_I(\xi) \mathbf{u}_I \tag{4.108}$$

where $\mathbf{P}_I$ are the control points that define the geometry and $\mathbf{u}_I$ are the control variables (analogous to nodal displacements in FEM, but associated with control points rather than nodes).

A distinctive feature of IGA is the higher inter-element continuity compared to traditional C^0 continuous Lagrange elements. B-spline/NURBS basis functions of degree p provide C^{p-m} continuity at knots with multiplicity m. This higher continuity is particularly advantageous for problems requiring higher-order derivatives, such as Kirchhoff-Love shells or gradient elasticity, eliminating the need for mixed formulations or special elements.

The strain-displacement relation follows the same form as in FEM:

$$\boldsymbol{\epsilon}^h(\xi) = \mathbf{B}(\xi)\mathbf{d} \tag{4.109}$$

where $\mathbf{B}$ contains the derivatives of the NURBS basis functions in physical coordinates. The computation requires the chain rule through the parametric mapping, similar to isoparametric finite elements.

The weak form discretization proceeds analogously to FEM, leading to the same linear system as in Eq. (4.84). The element stiffness matrix and force vector have the same structure as presented for FEM, but with NURBS basis functions replacing the Lagrange shape functions. However, the concept of "elements" in IGA differs from FEM—elements are defined as non-zero knot spans in the parameter space, while basis functions have support over multiple elements due to their higher continuity.

Integration is typically performed using Gauss quadrature in each knot span. Due to the higher polynomial degree and rational nature of NURBS, more quadrature points are often required compared to standard finite elements of equivalent polynomial order. Special quadrature rules have been developed to improve efficiency while maintaining accuracy.

A significant advantage of IGA is the exact representation of conic sections and other geometries commonly found in engineering designs, eliminating geometric approximation errors. The method also naturally handles trimmed geometries through trimming curves defined in parameter space. Additionally, refinement strategies in IGA (h-, p-, and k-refinement) preserve the exact geometry throughout the refinement process, with k-refinement being unique to IGA, allowing simultaneous increase in continuity and polynomial degree.

However, IGA faces several challenges that limit its widespread adoption. The tensor product structure of multivariate NURBS makes local refinement difficult, though techniques like T-splines, hierarchical B-splines, PHT-splines and LR-splines have been developed to address this limitation. The imposition of essential boundary conditions can be more complex than in FEM, as control points typically lie outside the physical domain and NURBS basis functions generally do not satisfy the Kronecker delta property (except at boundaries with open knot vectors). Furthermore, the tight integration with CAD systems requires significant changes to existing FEA software architectures, and many CAD models use boundary representations that require additional volumetric parameterization for solid analysis.

Despite these challenges, IGA has shown particular promise in applications involving complex geometries, shell and plate analysis, fluid-structure interaction, and shape optimization, where the seamless integration between design and analysis provides significant computational advantages. The method continues to evolve with ongoing research addressing its limitations while exploiting its unique advantages in maintaining exact geometry and higher continuity throughout the analysis process.

4.9.4 Comparison

ANNs are known to be universal approximators. Before focusing on some differences and similarities between FEM/IGA/MM solutions and a ML (or more precisely DNN) solution of PDEs, we would like to show the that DNNs using rectified linear units (ReLU) as activation functions can reproduce linear finite element spaces. This has been proven in [22], see also [23] for an extension to more general Finite Element spaces employing Sobolev norms. In this book, we adopt the proof from [22, 24] showing the equivalence of a DNN and linear FE interpolation in 1D. The DNN approximation is based on one hidden layer and employs ReLU functions defined by $ReLU(x) = \max(0, x)$. Let us consider a discretization of the interval $[0, l]$ in the real line: $x_0 < x_1 < \ldots < x_n$, where $x_0 = 0$ and $x_n = l$. Let us now consider a piecewise linear FE interpolation of a function $u(x)$ taking values on $[0, l]$:

$$u^h(x) = \sum_{I=1}^{n} N_I(x)\, \mathbf{u}_I \tag{4.110}$$

where $N_I(x)$ is a linear hat function with compact support in the subinterval $[x_{i-1}, x_{i+1}]$, such that $N_I(x_I) - 1$. For $I = 0$ and $I = n$, $N_I(x)$ has compact support in $[x_0, x_1]$ and $[x_{n-1}, x_n]$, respectively. Function $N_I(x)$, for intermediate nodes, can be expressed in terms of rectified linear units ReLU as follows:

$$N_I(x) = \frac{ReLU(x - x_{I-1}) - ReLU(x - x_I)}{h_I} - \frac{ReLU(x - x_I) - ReLU(x - x_{I+1})}{h_{I+1}} \tag{4.111}$$

with $h_I = x_I - x_{I-1}$ and $h_{I+1} = x_{I+1} h_I$. After some simple algebra, we can reformulate the shape functions to

$$N_I(x) = \frac{1}{h_I} ReLU(x - x_{I-1}) - \left[\frac{1}{h_I} + \frac{1}{h_{I+1}}\right] ReLU(x - x_I) + \frac{1}{h_{I+1}} ReLU(x - x_{I+1}) \tag{4.112}$$

Using Eqs. (4.112) and (4.110), the following expression can be derived:

$$u^h(x) = u_0 + \sum_{I=1}^{n-1} \left(\frac{\Delta_{I+1}}{h_{I+1}} - \frac{\Delta_I}{h_I} \right) ReLU(x - x_I) \tag{4.113}$$

with $\Delta_I = u_I - u_{I-1}, \forall I = 1, 2, \ldots, n$ and $\Delta_I = 0$ when $I = 0$. Now, u^h can be expressed as an ANN with one hidden layer with x as input, $u^h(x)$ as output, and ReLU as the activation function:

$$u^h(x) = u_0 + \sum_{I=1}^{n-1} w_I ReLU(x - x_I) \tag{4.114}$$

where $w_I = \left(\frac{\Delta_{I+1}}{h_{I+1}} - \frac{\Delta_I}{h_I} \right)$ are the weights and the position of the nodes x_I the biases of the hidden layer, which are commonly grouped in the vectors $\mathbf{w}$ and $\mathbf{x}_p$. Consider now the approximation in (4.113) and the associated 1-dimensional linear elasticity problem. Let us define $\psi(\epsilon)$ as the one-dimensional linear elastic strain energy density. Then, we obtain the associated energy by integrating over the length of the structure:

$$E(x) = \int_0^l \psi(\epsilon(u^h(x, \mathbf{w}, \mathbf{x}_p))) \tag{4.115}$$

In a deep energy method, the solution to this problem is formulated as optimization problem:

$$\underline{min}_{\mathbf{w},\mathbf{x}_p}\ E(\mathbf{w}, \mathbf{x}_p) \tag{4.116}$$

For fixed $\mathbf{x}_p$, the problem would be equivalent to solving a standard FE problem (remember the relationship between $\mathbf{w}$ and the nodal FE unknowns). When $\mathbf{x}_p$ is not fixed, from the mathematical point of view, we would be solving an adaptive FE problem. However, while in finite elements, we would solve a linear system of equations, in the deep energy method the problem is now non-linear and commonly non-convex, as pointed out in the introduction of this section. And though this 1D problem is illustrative concerning the meaning of the parameters of a DNN approximation, it is also clear that, for general situations, assigning physical meaning to these parameters would be very difficult. This may be a consequence of the mesh-free character of DNN approximations for the solution of PDEs. This has several implications. Collocation/quadrature/training points, for instance, are not to be confused with nodes in a Finite Element mesh. Points in a FE mesh are tightly related to the hyperparameters of the approximation space, while collocation points are the inputs for the training process in the DNN approximation. And while there are convergence proofs in finite elements, which guarantee—at least for specific problems—the convergence of the finite element method (meaning the results get more accurate with increasing mesh refinement), such proofs are not available in a DNN approximation, probably due to the richness of the approximation spaces. In fact, 'refining' the network by adding neurons and layers (which is comparable to h-refinement in finite elements) cannot, in general, guarantee a more accurate solution once issues like overfitting and the non-convexity of the loss function is taken into account. Furthermore, as in meshfree methods, there are several parameters to adjust including the number of hidden layer, number of neurons per layer and number of collocation/quadrature/training points which in turn will depend on the network architecture. Approaches based on optimization, sensitivity analysis and/or network architecture search are presented in Sect. 2.5.1.

Finally, here are some properties of the different computational approaches presented in this section compared to a machine learning based approach, see also Table 4.3:

- While FEM, IGA and MMs solve a linear or linearized system of equations, the ML model commonly solves nonlinear, non-convex optimization problems.
- While FEM approximations based on Lagrange polynomials are C^0 continuous, MMs, IGA and ML approximations are higher order continuous, which is advantageous e.g. for higher order PDEs. Note that the order of continuity in IGA increases 'only' when increasing the polynomial order, while even a linear complete MM can achieve C^∞. For instance, the continuity of EFG is governed by the continuity of the kernel function. Note also that while IGA approximations are higher order continuous within a single patch, the approximation is C^0 at patch boundaries. For complex geometries, it is indispensible to have multi-patches.
- FEM is interpolatory facilitating the imposition of essential boundary conditions while IGA, MMs and ML models are not interpolatory. Note that IGA fulfills the so-called weak Kronecker-delta property [21], such that an interpolation problem to enforce boundary conditions has to be solved only along the boundary.
- ML models scales better with increasing dimensionality compared to FEM, IGA and MMs.
- ML models are a natural framework for uncertainty analysis, optimization and inverse analysis. It can deal with such problems within the same framework.
- Concerning modeling complex geometries:
  - IGA (based on NURBS or similar rational splines) can model a wider range of geometries—such as conic cross sections—than Lagrange elements.
  - IGA has advantages in surface parametrizations but experiences difficulties in volume parametrizations due to the tensor product form.
  - The opposite applies to MMs: They are ideally suited for volume parametrizations. However, modeling complex surfaces/boundaries and microstructures is inherently difficult and might require techniques such as the visibility method [17].
  - ML models can exploit any geometry parametrization (based on FEM, IGA, MMs, etc.).

Table 4.3 Comparison of computational methods for solving PDEs

Property	FEM	IGA	MM (EFG)	ML-based
Problem type	Linear or linearized system of equations	Linear or linearized system of equations	Linear or linearized system of equations	Nonlinear, commonly non-convex optimization problem
Continuity	C^0 (Lagrange polynomials)	Higher-order continuous within patch (C^{p-m} for degree p, knot multiplicity m); C^0 at patch boundaries	C^∞ possible (governed by kernel function continuity)	Higher-order continuous
Interpolatory	Yes (fulfills Kronecker delta property)	Weak Kronecker delta property at boundaries	No	No
BC imposition	Straightforward (strong form)	Requires boundary interpolation only	Complex (requires special techniques)	Complex
Convergence guarantees	Yes (certified solution with mesh refinement)	Yes	Yes (for specific cases)	No (due to non-convexity, overfitting risks)
High dimensionality scaling	Poor	Poor	Poor	Better
Complex geometries	Limited (polynomial approximation)	Excellent for surfaces (exact conics); difficult for volumes (tensor product)	Excellent for volumes; difficult for complex surfaces	Can exploit any parametrization (FEM/IGA/MM)
Local refinement	Straightforward	Difficult (tensor product structure); addressed by T-splines, hierarchical B-splines	Natural (variable support sizes)	Flexible (architecture adaptation)
Integration/ Quadrature	Gauss quadrature (background mesh)	Gauss quadrature (more points due to rational functions)	Gauss quadrature or background mesh	Training/collocation points (not tied to mesh)
Mesh dependency	Requires mesh	Requires parametric mesh	Requires background mesh (for integration)	Mesh-free
Uncertainty & inverse analysis	Requires separate framework	Requires separate framework	Requires separate framework	Natural framework (built-in)

(continued)

Table 4.3 (continued)

Property	FEM	IGA	MM (EFG)	ML-based
Calibration complexity	Well-established (element size, order)	Moderate (knot vector, weights)	Moderate to high (support size, kernel parameters)	High (layers, neurons, training points, architecture)
Nodal parameters	Physical values at nodes	Control variables at control points	Non-physical (approximation coefficients)	Network weights and biases (non-physical)
Computational cost (setup)	Low to moderate	Moderate	Higher (neighbor search, shape function computation)	High (training process)
Best suited for	General linear problems, well-posed BVPs	Complex CAD geometries, design-analysis integration	Large deformations, fracture, fragmentation	Ill-posed problems, high-dimensional problems, inverse problems, optimization

References

1. Raissi, M., Perdikaris, P., Karniadakis, G.E.: Physics-informed neural networks: A deep learning framework for solving forward and inverse problems involving nonlinear partial differential equations. J. Comput. Phys. **378**, 686–707 (2019). 10.1016/j.jcp.2018.10.045
2. E, W., Bing, Y.: The deep Ritz method: a deep learning-based numerical algorithm for solving variational problems. Commun. Math. Stat. **6**(1), 1–12 (2018). 10.1007/s40304-018-0127-z
3. Hager, W.W., Zhang, H.: Algorithm 851: CG_DESCENT, a conjugate gradient method with guaranteed descent. ACM Trans. Math. Softw. (TOMS) **32**(1), 113–137 (2006). 10.1145/1132973.1132979
4. Dillon, J.V., Langmore, I., Tran, D., Brevdo, E., Vasudevan, S., Moore, D., et al.: TensorFlow distributions. Preprint. arXiv:1711.10604 (2017). 10.48550/arXiv.1711.10604
5. Wang, S., Teng, Y., Perdikaris, P.: Understanding and mitigating gradient flow pathologies in physics-informed neural networks. SIAM J. Sci. Comput. **43**(5), A3055–A3081 (2021). 10.1137/20M1318043
6. Wang, S., Wang, H., Perdikaris, P.: On the eigenvector bias of Fourier feature networks: From regression to solving multi-scale PDEs with physics-informed neural networks. Comput. Methods Appl. Mech. Eng. **384**, 113938 (2021). 0045-7825. 10.1016/j.cma.2021.113938
7. Wang, S., Yu, X., Perdikaris, P.: When and why PINNs fail to train: A neural tangent kernel perspective. J. Comput. Phys. **449**, 110768 (2022). 10.1016/j.jcp.2021.110768
8. Wang, Y., Sun, J., Li, W., Lu, Z., Liu, Y.: CENN: Conservative energy method based on neural networks with subdomains for solving variational problems involving heterogeneous and complex geometries. Comput. Methods Appl. Mech. Eng. **400**, 115491 (2022). 10.1016/j.cma.2022.115491
9. Ghia, U., Ghia, K.N., Shin, C.T.: High-Resolutions for incompressible flow using the Navier-Stokes equations and a multigrid method. J. Comput. Phys. **48**(3), 387–411 (1982)
10. Schillinger, D., Borden, M.J., Stolarski, H.K.: Isogeometric collocation for phase-field fracture models. Comput. Methods Appl. Mech. Eng. **284**, 583–610 (2015). 10.1016/j.cma.2014.09.032
11. Miehe, C., Hofacker, M., Welschinger, F.: A phase field model for rate-independent crack propagation:Robust algorithmic implementation based on operator splits. Comput. Methods Appl. Mech. Eng. **199**(45–48), 2765–2778 (2010). 10.1016/j.cma.2010.04.011

12. Bittencourt, T., Wawrzynek, P., Ingraffea, A., Sousa, J.: Quasi-automatic simulation of crack propagation for 2D LEFM problems. Eng. Fract. Mech. **55**(2), 321–334 (1996). 10.1016/0013-7944(95)00247-2
13. Molnár, G., Gravouil, A.: 2D and 3D Abaqus implementation of a robust staggered phase-field solution for modeling brittle fracture. Finite Elem. Anal. Des. **130**, 27–38 (2017). 10.1016/j.finel.2017.03.002
14. Hirshikesh, Natarajan, S., Annabattula, R.K.: A FEniCS implementation of the phase field method for quasi-static brittle fracture. Front. Struct. Civil Eng. **13**(2), 380–396 (2019). 10.1007/s11709-018-0471-9
15. Fernández-Méndez, S., Huerta, A.: Imposing essential boundary conditions in mesh-free methods. Comput. Methods Appl. Mech. Eng. **193**(12–14), 1257–1275 (2004). 10.1016/j.cma.2003.12.019
16. Rabczuk, T., Zi, G.: A meshfree method based on the local partition of unity for cohesive cracks. Comput. Mech. **39**(6), 743–760 (2007). 10.1007/s00466-006-0067-4
17. Belytschko, T., Lu, Y.Y., Gu, L.: Element-free Galerkin methods. Int. J. Numer. Methods Eng. **37**(2), 229–256 (1994). 10.1002/nme.1620370205
18. Liu, W.K., Jun, S., Zhang, Y.F.: Reproducing kernel particle methods. Int. J. Numer. Methods Fluids **20**(8–9), 1081–1106 (1995). 10.1002/fld.1650200824
19. Belytschko, T., Krongauz, Y., Organ, D., Fleming, M., Krysl, P.: Meshless methods: an overview and recent developments. Comput. Methods Appl. Mech. Eng. **139**(1–4), 3–47 (1996). 10.1016/S0045-7825(96)01078-X
20. Li, S., Liu, W.K.: Meshfree and particle methods and their applications. Appl. Mech. Rev. **55**(1), 1–34 (2002). 10.1115/1.1431547
21. Hughes, T.J., Cottrell, J.A., Bazilevs, Y.: Isogeometric analysis: CAD, finite elements, NURBS, exact geometry and mesh refinement. Comput. Methods Appl. Mech. Eng. **194**(39–41), 4135–4195 (2005). 10.1016/j.cma.2004.10.008
22. He, J., Li, L., Xu, J., Zheng, C.: ReLU deep neural networks and linear finite elements. Preprint. arXiv:1807.03973 (2018). 10.4208/jcm.1901-m2018-0160
23. Opschoor, J.A., Petersen, P.C., Schwab, C.: Deep ReLU networks and highorder finite element methods. Anal. Appl. **18**(05), 715–770 (2020). 10.1142/S0219530519410136
24. Samaniego, E., Anitescu, C., Goswami, S., Nguyen-Thanh, V.M., Guo, H., Hamdia, K., et al.: An energy approach to the solution of partial differential equations in computational mechanics via machine learning: Concepts, implementation and applications. Comput. Methods Appl. Mech. Eng. **362**, 112790 (2020). 10.1016/j.cma.2019.112790

Chapter 5
Surrogate Models Based on Machine Learning

5.1 Machine Learning Surrogate Models in Computational Mechanics

The development of computational mechanics makes it possible to analyze various complex phenomena. For the high-fidelity model, the prediction of its behavior requires costly and time-consuming simulation. Especially, certain applications of practical interest such as optimization, uncertainty quantification, and parameter identification require a large number of model runs.

The surrogate model is a type of approximation model, which is developed to deal with black-box problems and lower computational requirements. Machine learning has become a popular surrogate modeling approach and has shown great potential to deal with situations where input-output mappings are extensively complex. In addition, machine learning methods can be employed to excavate inherent rules and correlations based on a significant amount of data samples [1].

The key point is to exploit the features of a particular machine learning model for specific problems in computational mechanics. For example, the feedforward neural network (FNN) can build mapping relationships in a general sense. The recurrent neural network (RNN) is suitable for handling time series databases, so it is often used for the path or history-dependent problems. The convolutional neural network (CNN) is commonly applied to analyze visual imagery and is utilized in model reduction and material parameter identification. The generative adversarial network (GAN) is often used in combination with the above network to expand the dataset. Of course, other less common machine learning models and the combination of the above models are also applied in computational mechanics. The simplest machine learning model is FNN, and it is also the most common model in ML-based computational mechanics. This is because it simply establishes mapping relationships, and there are no restrictions on the application scenarios.

T. Rabczuk et al., *Scientific Machine Learning with Engineering Applications*,
Studies in Systems, Decision and Control 667,
https://doi.org/10.1007/978-3-032-20307-6_5

Correspondingly, models with distinct features, such as RNN/CNN/GAN, are suitable for a specific class of problems.

Replacement or integration is a central issue in the application of machine learning to computational mechanics. The seamless integration approach of ML and simulation mostly employs the FNN model, but a suitable entry point is difficult to find, thus relevant research is still limited. The integration strategy retains advantages of both, namely the generality of the numerical methods and the low computational cost of machine learning. Such methods are usually data-free. In the strategy of replacing numerical simulation with machine learning, the database is generated with numerical methods for the training of surrogate model. Due to the limitations in the dataset or the methodology, this approach is applicable to a particular problem, and its generalization is limited. Machine learning can make predictions in a short time, but the dataset generation and design consume a lot of time.

5.2 Applications of Surrogate Models

5.2.1 Multiscale and Constitutive Model

The hierarchical multiscale method based on the representative volume element (RVE) is a type of multiscale method that offers the numerical constitutive relationship at the macroscopic point depending on the morphology and evolution of the microscale model. It is similar to the simulation of the traditional material constitutive model.

Different strategies for multiscale and constitutive models based on machine learning are given in Table 5.1. The most popular neural networks are FNN, RNN, and CNN. In addition, GAN and combinations of different neural networks are also used in multiscale modeling.

FNN is the simplest and most intuitive neural network, and its applications are diverse. FNN can be used to construct the averaged strain energy density of RVE [2, 3], and also been employed to approximate constitutive behavior by fitting the stress-strain relation directly [4], where the damage information can also be included [5] (Fig. 5.1). Another strategy to reduce the computation cost with FNN is to build a deep material network for model reduction [6, 7], where RVE is represented by a hierarchical topological structure with mechanistic building blocks. The nonlinear material behaviors can be predicted both accurately and efficiently (Fig. 5.2).

RNN is an extension of the neural network designed to handle sequential data, which is used to account for history-dependent behaviors. To train the surrogate, one needs to collect a set of strain tensor sequences and their corresponding stress tensor sequences from the micromodel. However, the space of all possible strain sequences is extremely large and sampling is therefore computationally intractable. The sampling strategies in the literature are listed in Table 5.2.

Table 5.1 Different strategies for multiscale and constitutive model based on machine learning

Constitutive	NN model	Highlight	Reference
Hyperelastic	FNN	Approximation of homogenized energy density	Minh [2], Linka [3]
Elastoplastic	FNN	Path-independent, mapping database from FE^2	Xu [4]
Elastoplastic, damage	FNN	Homogenized stress-strain relationship and damage information with UMAT	Yan [5]
Elastoplastic	FNN	A deep material network for model reduction	Liu [6, 7]
Elastoplastic	RNN	History-dependent microlevel response, crossscale simulation	Mozaffar [8], Abueidda [9], Ghavamian [10], Wu [11]
Elastic, discontinuity	RNN	A multiscale multipermeability poroplasticity model (Fig. 5.3)	Wang [12]
Elastic	CNN, TL	Mapping structure-property in multiscale model (Fig. 5.4)	Yang [13], Wei [14], Vasilyeva [15], Rao [16]
AZS composite	GAN, MSG	Generate microstructure using GAN, database of structure genome	Samaei [17]
Aluminum crystal	FNN, SVM	Material failure classification, stress regression	Xiao [18]
Crystal plasticity,	CNN, RNN	A hybrid CNN-RNN network for the response of oligocrystals	Frankel [19]
Elastoplastic	GP	Adaptive probabilistic ML, on-the-fly, monotonic loading	Rocha [20]

TL: transfer learning; GP: Gaussian process; MSG: mechanics of structure genome; SVM: support vector machine

Using artificial neural network (FNN or RNN) to approximate nonlinear constitutive behavior by fitting the stress-strain directly is usually problem-dependent and may suffer from "the danger of extrapolation" beyond the original sampling space, e.g. different material laws and loading paths. Issues like material history dependency, physical invariance, and conservation laws are not naturally resolved, mainly due to the loss of physics in the current machine learning models [6].

The consistent tangent can be computed using the deep learning library's automatic differentiation functionality:

$$\boldsymbol{D}^t = \frac{\partial \boldsymbol{\sigma}^t}{\partial \boldsymbol{\varepsilon}^t} \tag{5.1}$$

CNN-based approach is usually developed as a highly efficient and portable tool to identify the elastic isotropy of architected materials directly from images of the unit cells with arbitrary component distributions. Moreover, transfer learning

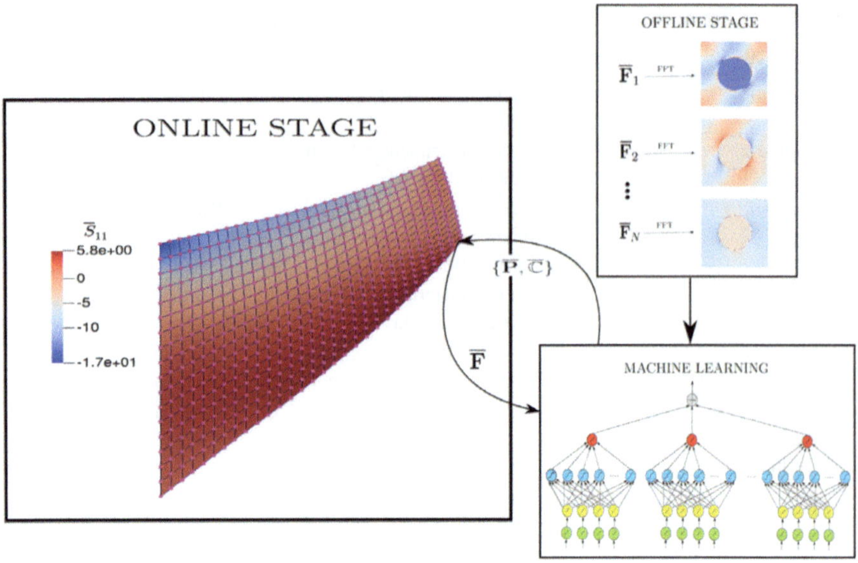

Fig. 5.1 Surrogate model for computational homogenization by means of approximator of macroenergy density [2]

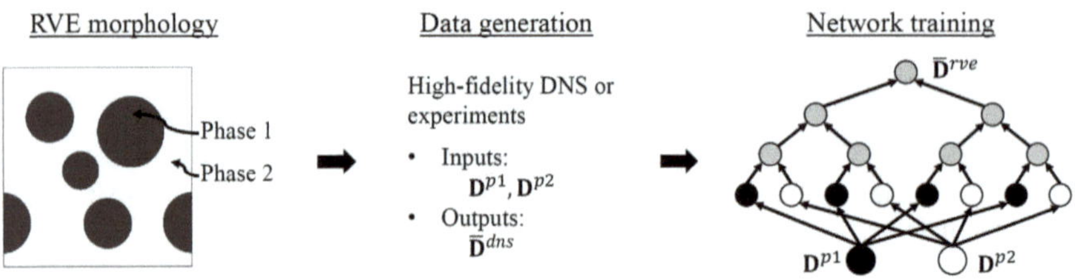

Fig. 5.2 Training procedure of material reduction network [6]

Table 5.2 Different sampling strategies for RNN based multiscale modelling

Data set	Input-output	Reference
15,000 different deformation paths	Strain-stress, plastic energy	Mozaffar [8]
Random strain on control points, 15,500 datapoints	Strain-stress, plastic energy	Abueidda [9]
Possible strain sequence from a set of macro models	Strain-stress	Ghavamian [10]
Around 9000 of random loading path	Strain-stress	Wu [11]

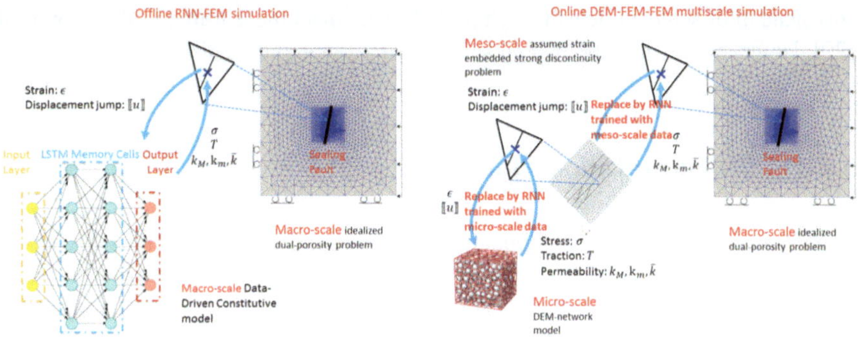

Fig. 5.3 Comparison between off-line pre-trained multiscale ANN-FEM simulations and online hierarchical multiscale simulations [12]

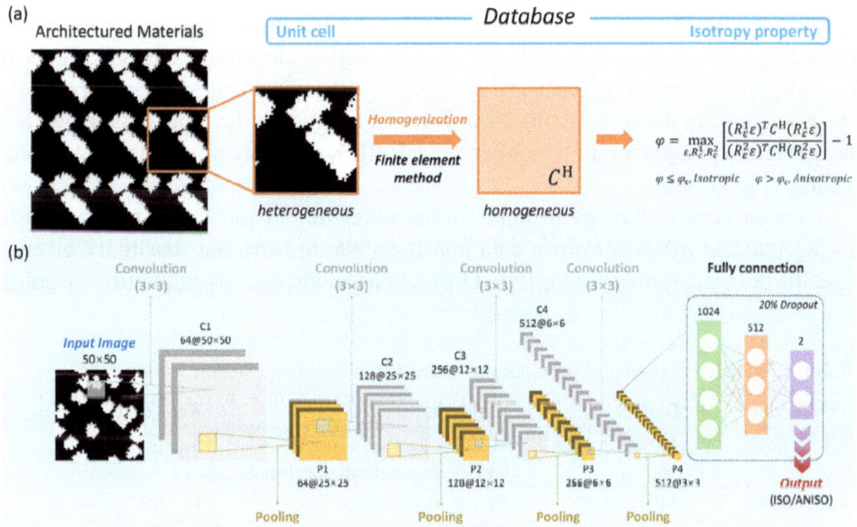

Fig. 5.4 Implementation of the deep learning-based approach for the isotropy identification of architected materials [14]

(TL) is often implemented among architectured materials with different numbers of material components, broadening the application scope of this method.

An issue for CNN models in predicting the effective elastic properties is the sufficient database construction of microstructure, options include microscale volume elements (MVES) [13], high-resolution tomography [15], Hierarchical Random Sequential Adsorption (HRSA) algorithm [16] and so on. GAN is used to generate microstructure with varying porosities, complex pore shapes, and randomly distributed pores [17].

In addition, different types of neural networks can be combined to achieve more functions [18, 19]. Without an additional dataset, on-the-fly construction of surrogate multiscale models through probabilistic machine learning is also proposed [20]. However, this method is only suitable for monotonic loading paths.

5.2.2 Structural Optimization

Topology optimization is computationally expensive due to the many required iterations. Generally, the use of machine learning for topology optimization can be divided into three types, i.e. the heuristic algorithm (CNN, GAN etc.), the surrogate model from iterative history data, and present density field presented with FNN, as shown in Table 5.3.

Accelerating topology optimization with CNN is a gradient-free method, which is easy to implement. However, it requires a relatively large dataset. Moreover, this method is only for some specific problems, and it can not take effect when the boundary conditions, objective function, or material constitutive changes. GAN is often used to learn the geometrical features of the input images and generate new images that have the same geometrical features as those in the input data. However, the generated images might be similar to each other, therefore providing only a small variety (Fig. 5.5).

Another typical strategy to improve the efficiency of topology optimization is to dig into the iterative history data and use FNN to learn the sensitivity given a specific design, as shown in Fig. 5.6. Different methods, such as calibration by going

Table 5.3 Different strategies for topology optimization based on machine learning

NN model	Dataset	Highlight	Reference
CNN, GAN	44,000 images	Model reduction using CNN, Satisfy geometrical constraints	Tan [21]
CNN, GAN	1407 airfoils	Aerodynamic shape optimization for airfoils and wings	Li [22]
CNN, GAN, TL	16,689 wheel designs	A deep learning-based CAD/CAE framework for the conceptual design	Yoo [23]
CNN, GA	2,000,000 points	Accuracy of the surrogate model	Messner [24]
CNN (ResUnet)	15,000–30,000 pairs	Parallel schemes, nonlinearities; metamaterial design	Abueidda [25], Kollmann [26]
CNN, FNN	614,304 (2D)	Three stage NN based topology optimization approach	Bielecki [27]
FNN/DBN	Iterative history data	Training during the TO; parallel schemes; tailored two-scale TO formulation	Chi [28], Bi [29], Kallioras [30]
FNN	Without dataset	Density distribution is characterized by NN	Deng [31], Chandrasekhar [32]

DBN: Deep belief network; GA: Genetic algorithm

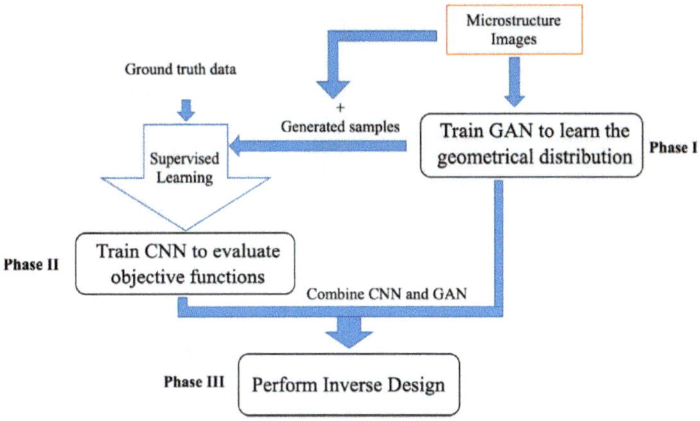

Fig. 5.5 Flowchart diagram for topology optimization with GAN and CNN [21]

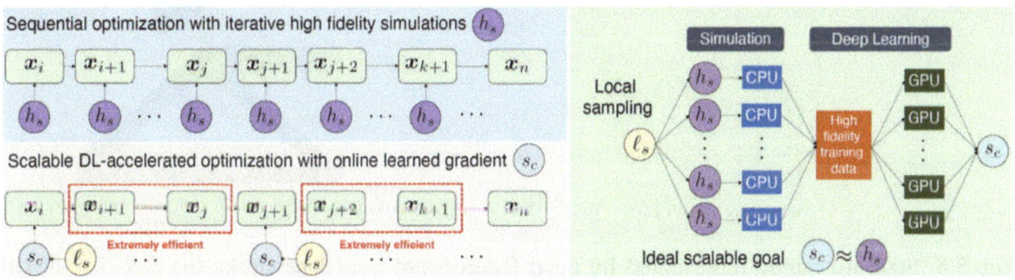

Fig. 5.6 Scalable implementation of TO by learning the Iterative history data with multiple CPUs and GPUs [29]

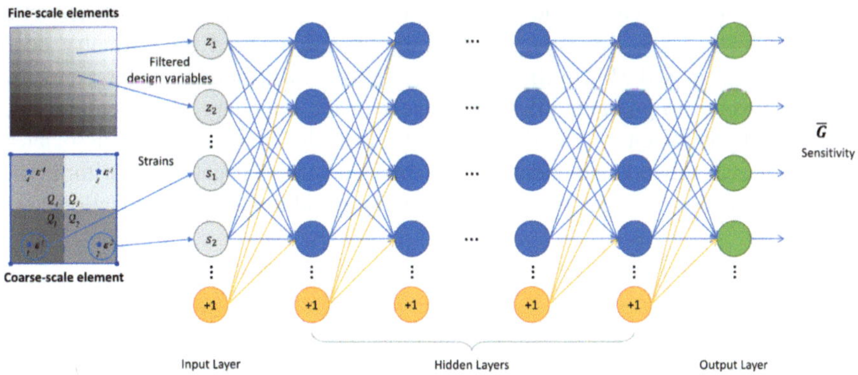

Fig. 5.7 Architecture of the fine-coarse mesh topology optimization setup with DNN [28]

back to topology optimization at the end [30], a tailored fine-coarse formulation [28] (Fig. 5.7), proper setup for training and update step [28, 29], are employed to enhance the performance of the method.

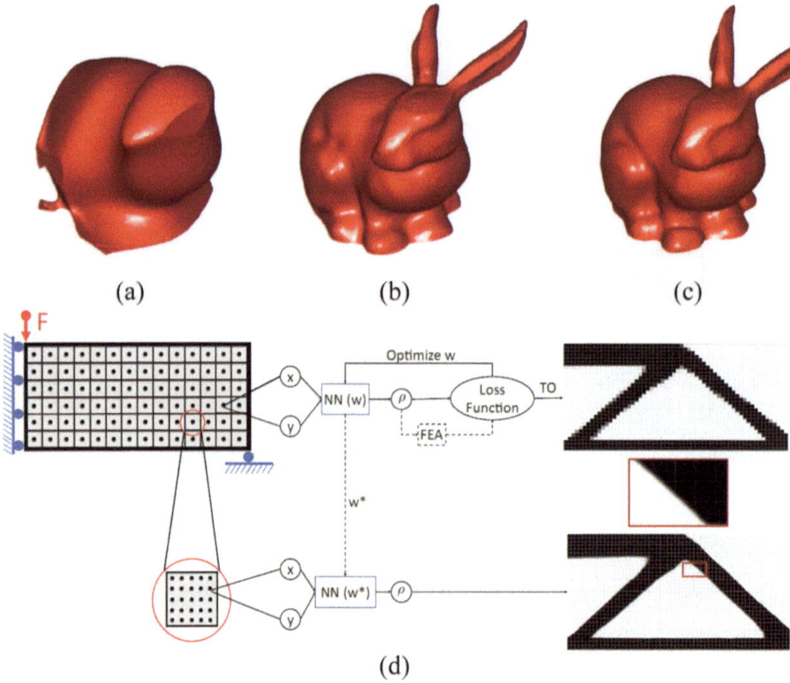

Fig. 5.8 Stanford bunny represented by deep feedforward neural networks (**a**) $5 \times 5 \times 5$, (**b**) $20 \times 20 \times 20$, (**c**) $40 \times 40 \times 40$ [31]; (**d**) Topology Optimization using Neural Networks to present density field [32]

Complex models can be described by the relatively shallow neural network, as shown in Fig. 5.8. Recently, topology optimization based on geometry deep learning is proposed (Fig. 5.8), where the density distribution is described by neural networks to guarantee the smoothness of the boundary and overcome the checkboard phenomenon. The design variables are the weight and bias associated with the neural network and updated by the built-in backpropagation algorithm. The method is a real "marriage" between deep learning and topology optimization.

5.2.3 PDE

In conventional ML problems, the mechanism behind the system is usually unknown and thus can only be learned from the labeled data. Similarly, a type of surrogate model for PDE uses label data to map the problem's parameter space to its solution space [33–35]. Instead of learning solely from the labeled data, the known governing equations can be utilized to constrain the learning to compensate for the insufficiency of the data (Table 5.4).

Recently, the so-called physics-informed neural networks (PINN) or physics-constrained neural networks (PCNN) have been proposed to learn and encode

Table 5.4 Data-driven surrogate modeling and physics-constrained learning for PDE problems

NN model	Equation	Highlight	Reference
FNN	Euler-Bernoulli beam	Two sub-networks to approximate the shape function, displacement and velocity response	Wang [33]
PCA, CNN	Maxwell's equation	Predict subspace for augmenting the solver iteration	Trivedi [34]
CAEs, FNN	Parametrized time-dependent PDEs	Mapping from the problem's parametric space to its solution space	Nikolopoulos [35]
PCNN	Laplace differential equations	Label-free, add simulation data or labels at collocation points	Jin [36]
CAEs	Parametric PDEs	Label-free, uncertainty quantification	Zhu [37]
FC-NN	Parametric N-S equations	Label-free, hard boundary enforcement	Sun [38]

PCNN: physics-constrained neural network; CAEs: Convolutional Autoencoders; FC-NN: PDE-constrained fully-connected neural network

physics laws expressed by nonlinear PDE [36], as shown in Fig. 5.9. PINN/PCNN promises that only a small number of training examples are needed, and the resulting DNN models will quickly reflect the underlying principled physics laws for all the input coordinates and parameters so that fast and accurate numerical solutions can be obtained.

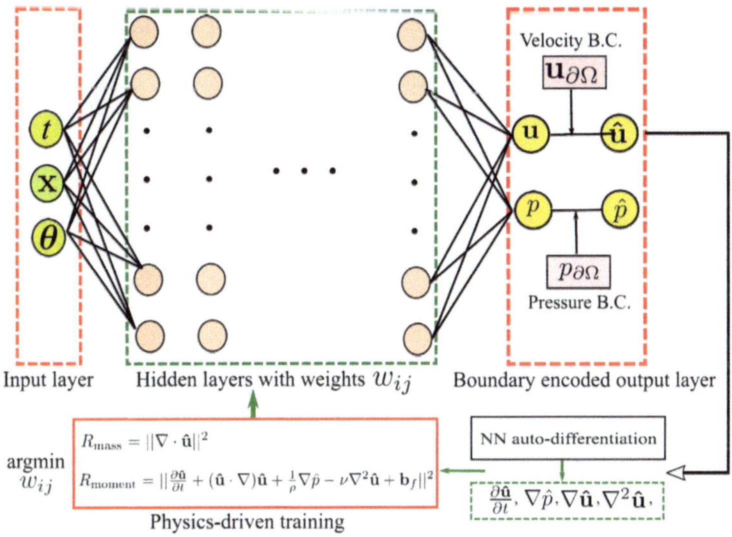

Fig. 5.9 A schematic diagram of the physics-constrained, data-free DL framework for surrogate modeling of fluid flows [38]

5.2.4 Improvement of FEM Solutions

The strategies for the improvement of FEM solutions generally can be classified into two categories. A common approach is to replace the FEM with machine learning. The database is designed with results from numerical simulations and used to train the data-driven black box model (Table 5.5). Related applications include behaviour prediction [40, 43, 45], inverse analyse [41, 44], uncertainty propagation [39], data assimilation [42] and dynamic response prediction [46–49].

A more challenging task is to integrate machine learning with FEM to make them a combined universal method. Examples include interpolation functions [50], numerical quadrature [51] and substructuring technique [52].

Table 5.5 Different strategies for the improvement of FEM solutions based on machine learning

NN model	Objective	Highlight	Reference
CNN	Uncertainty propagation	Characterize the system responses under various spatially varying properties	Luo [39]
FNN, PCA	Substitute FEM	Estimate stress distribution and inverse analyze for the human thoracic aorta	Liang [40, 41]
DNN	Detection and classification	Data assimilation, improve the FEM of the metal cutting process	Misaka [42]
FNN	Behaviour prediction	Surrogate model for carbon nanotubes with geometric nonlinearities	Papadopoulos [43]
PCA, FNN	Identify load conditions	Inverse solution to determine the impact load conditions of shell structures	Chen [44]
CNN, RNN	Predict stress	Accurate multi-step predictions of stress with fracture propagation in brittle materials (Fig. 5.10)	Wang [45]
RNN	Dynamic response prediction	A series of time sequence problems in civil engineering	Ninic [46], Koeppe [47], Cao [48], Peng [49]
HiDeNN	Interpolation functions	Produces a global shape function with compact support with DNN (Fig. 5.11)	Zhang [50]
FNN	Numerical quadrature	A new optimized numerical quadrature enhanced by deep learning	Oishi [51]
TRUNet	Model reduction	A nonlinear intelligent meta element using substructuring technique (Fig. 5.12)	Koeppe [52]

TRUNet: The time-distributed residual u-net; HiDeNN: The hierarchical deep-learning neural network

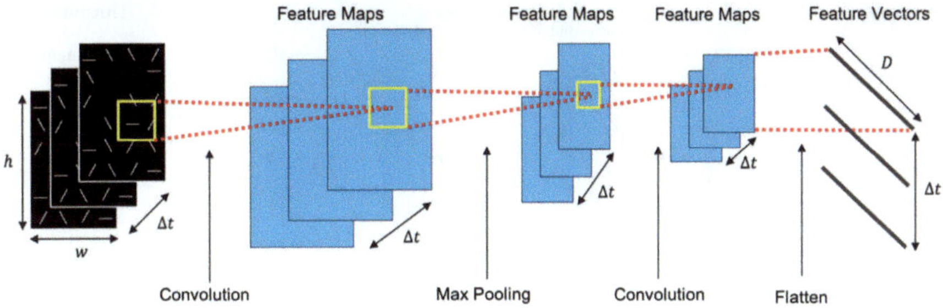

Fig. 5.10 Architecture of the RNN-CNN model for fracture propagation [45]

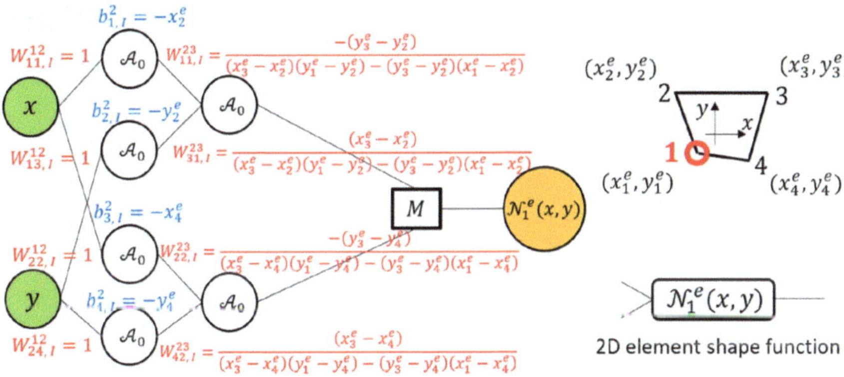

Fig. 5.11 DNN representation for 2D element shape function [50]

5.2.5 *Flow Simulation*

The application of machine learning to computational fluid dynamics (CFD) has emerged as a transformative approach to accelerate simulations, enhance accuracy, and enable real-time predictions. Various deep learning architectures have been developed to address different aspects of flow simulation, each leveraging specific neural network characteristics to tackle the inherent challenges in fluid mechanics (Table 5.6).

5.2.5.1 Data-Driven Approaches for Flow Prediction

Feedforward neural networks (FNN) serve as the foundation for establishing direct mapping relationships between flow parameters and field variables. They are particularly effective for steady-state flow problems and parameter studies where the relationship between inputs (geometry, boundary conditions, material properties) and outputs (velocity, pressure, temperature fields) can be learned from simulation data. Recent work has demonstrated that FNNs can capture complex nonlinear

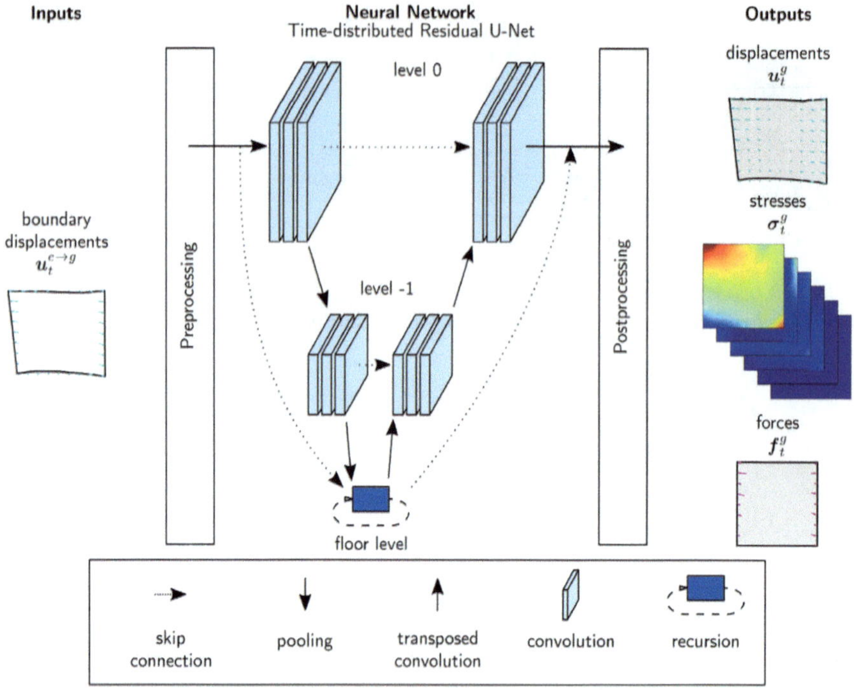

Fig. 5.12 The proposed intelligent nonlinear meta element with substructuring technique [52]

flow behaviors in multi-scale frameworks [53] and multiphase flow simulations [54], though they typically require extensive training datasets generated from high-fidelity simulations.

Recurrent neural networks (RNN) and their variants, such as Long Short-Term Memory (LSTM) networks, have proven particularly valuable for time-dependent flow problems. These architectures naturally incorporate temporal dependencies, making them suitable for unsteady aerodynamics [55, 56] and dynamic flow evolution. The sequential nature of RNNs allows them to learn the temporal evolution of flow fields, capturing transient phenomena and history-dependent behaviors. Generative adversarial networks (GAN) have been combined with RNNs to generate high-fidelity turbulent flow realizations [57], addressing the challenge of producing physically realistic turbulent structures.

Convolutional neural networks (CNN) excel at processing spatial information and have become a cornerstone for flow field prediction. Their ability to extract hierarchical features from structured data makes them ideal for learning spatial patterns in velocity, pressure, and other field variables. CNNs have been successfully applied to accelerate Reynolds-Averaged Navier-Stokes (RANS) simulations [58], predict steady laminar flows in complex geometries [59], and model flow in porous media [60]. The U-Net architecture and its variants, particularly R-U-Net (Fig. 5.13) incorporating residual connections and recurrent layers [61–63], have shown

Table 5.6 Different strategies for CFD simulation based on machine learning

NN model	Objective	Highlight	Reference
FNN	Multi-scale framework	Prediction of bubbly flow using physics-guided deep learning	Bao [53]
FNN	Multiphase flow simulation	Approximate flow and transport equations	Wang [54]
RNN	Airfoil–gust interaction	Dataset generated based reduced order model	Halder [55]
RNN	Aeroelastic response	Predict aerodynamic force and moment coefficients based on motion history	Srivastava [56]
RNN, GAN	High-fidelity turbulent flow	Generate physically accurate turbulent realizations	Geneva [57]
CNN	Accelerate the coverage	Accelerating the Reynolds Averaged Navier-Stokes simulations (Fig. 5.14)	Obiols [58]
CNN	Steady flow approximation	Real-time prediction of non-uniform steady laminar flow in a 2D or 3D domain	Guo [59]
U-ResNet	Porous media	Captures the flow fields at the grid level for samples	Takbiri [60]
R-U-Net	Dynamic subsurface flow	Evolution of pressure and saturation	Tang [61–63]
CNN	Complex environments	Scalable CNN for fluid flow prediction	Rana [64]
PINN	Thermal cavity flow	Natural convection with vorticity-stream formulation	Fowler [65]
PINN	Turbulent flows	RANS-based PINNs with turbulence models	Pioch [66]
PINN	Multiphase flow	Buckley-Leverett equation with shock handling	Zhang [67]
PINN	Shallow water	Label-free training for 2D shallow water equations	Qi [68]
PINN	Flow optimization	Adaptive sampling and loss weighting	Zhou [69]
DNN	Turbulence forecasting	Cross-Reynolds number prediction	Buaria [70]
GAN	3D reconstruction	Reconstruct 3D turbulent fields from 2D data	Yousif [71]
MeshGraphNet	Large-scale CFD	Scalable GNNs for unstructured meshes	Lei [72]
FNO	Operator learning	Fourier Neural Operators for turbulent flows	Li [73]
DRL	Active control	Reinforcement learning for turbulent separation	Font [74]
DBN	Design optimization	Multi-fidelity surrogate model to approach the CFD results	Tao [75]

DBN: deep belief network; PINN: physics-informed neural network; DRL: deep reinforcement learning; FNO: Fourier Neural Operator; GNN: graph neural network

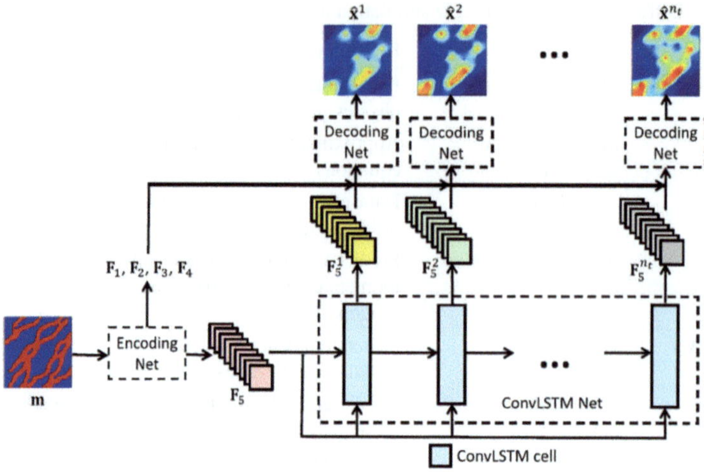

Fig. 5.13 Recurrent R-U-Net architecture incorporating convLSTM into the R-U-Net [61]

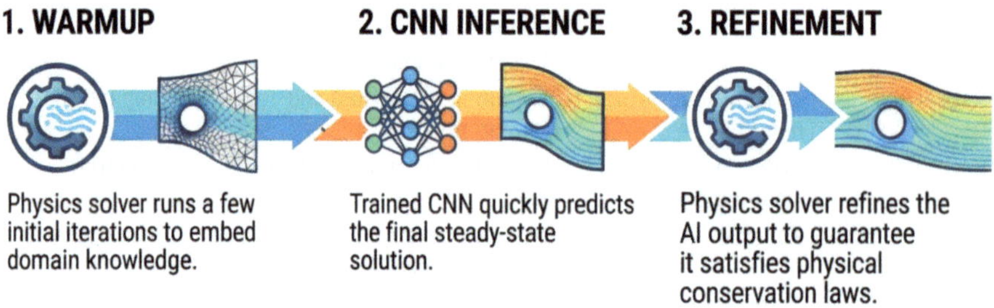

Fig. 5.14 Integration of the traditional physics solver simulation with CFDNet, adapted from [58]

remarkable performance in predicting dynamic subsurface flow evolution, capturing both pressure and saturation fields with high accuracy.

5.2.5.2 Physics-Informed Neural Networks for Flow Simulation

Physics-informed neural networks (PINNs) represent a paradigm shift in machine learning for CFD by embedding governing equations directly into the neural network training process. Unlike purely data-driven approaches, PINNs incorporate the Navier-Stokes equations, conservation laws, and boundary conditions as soft constraints in the loss function, enabling training with minimal or no labeled data [38, 76]. This physics-informed approach offers several advantages: reduced data requirements, better extrapolation capabilities, and guaranteed physical consistency.

Recent advances in PINNs have addressed several key challenges. Adaptive sampling strategies that concentrate training points in regions with high equation residuals have significantly improved accuracy [69]. The application of PINNs to turbulent flows requires incorporating turbulence models, with recent work demonstrating successful implementation of RANS-based PINNs for backward-facing step flows and other benchmark cases [66]. For thermal convection problems, PINNs using the vorticity-stream function formulation of the Navier-Stokes equations have shown promising results [65]. However, PINNs typically require 3–5 times more computational time than traditional CFD during training, though they offer 5–10 times reduction in memory usage and enable rapid inference for new parameter configurations [77].

The extension of PINNs to multiphase flows presents additional challenges due to discontinuities in the solution. Recent work has successfully applied PINNs to the Buckley-Leverett equation for modeling fluid flow in porous media, achieving superior precision by incorporating Welge's construction to handle shock discontinuities [67]. For free surface flows governed by shallow water equations, PINNs eliminate the need for labeled training data while maintaining accuracy in capturing complex flow dynamics [68].

5.2.5.3 Graph Neural Networks and Operator Learning

Graph neural networks (GNNs) have emerged as a powerful tool for CFD simulation due to their ability to handle unstructured meshes and irregular geometries. MeshGraphNets represent flow fields as graphs where nodes correspond to mesh elements and edges encode spatial relationships [78]. This representation naturally accommodates mesh refinement, complex boundaries, and dynamic mesh adaptation. Recent developments have scaled GNN training to meshes with millions of nodes, enabling application to industrial-scale problems [72].

Operator learning approaches, including Fourier Neural Operators (FNO) and DeepONet, learn mappings between infinite-dimensional function spaces rather than finite-dimensional vectors. These architectures can predict entire solution fields for new parameter configurations without retraining, offering superior generalization compared to traditional neural networks. FNOs have demonstrated remarkable performance for turbulent flow prediction, learning the solution operator that maps initial conditions and parameters to flow evolution [73].

5.2.5.4 Turbulence Modeling with Deep Learning

Turbulence modeling remains one of the most challenging aspects of CFD, and machine learning offers new opportunities to improve turbulence closures. Deep neural networks have been trained to forecast small-scale turbulent dynamics, with studies demonstrating that networks trained on lower Reynolds number simulations can successfully predict turbulent structures at higher, unseen Reynolds numbers [70]. This cross-Reynolds number generalization capability represents a significant advancement in turbulence modeling.

Generative models, particularly GANs, have been employed to reconstruct three-dimensional turbulent velocity fields from two-dimensional observations [71], addressing a longstanding challenge in experimental fluid mechanics. These models learn the underlying turbulent structures and statistics, enabling high-fidelity reconstruction of unmeasured flow components. Deep reinforcement learning has also been applied to active flow control in turbulent separation, learning optimal control strategies that manipulate flow structures to reduce drag or enhance mixing [74].

5.2.5.5 Hybrid Approaches and Industrial Applications

Hybrid approaches that combine traditional CFD solvers with machine learning components offer a practical path to leveraging ML advantages while maintaining physical rigor. CNNs can be used to accelerate the convergence of iterative solvers by predicting good initial guesses or correction fields [58]. Multi-fidelity surrogate models that combine low-fidelity simulations with ML corrections enable design optimization workflows where high-fidelity CFD evaluations are prohibitively expensive [75].

Industrial applications increasingly employ deep belief networks (DBN) and ensemble methods for robust flow prediction across diverse operating conditions. The integration of transfer learning allows models trained on one flow regime or geometry to be adapted to related problems with minimal additional training data. However, challenges remain in ensuring model robustness, quantifying prediction uncertainty, and maintaining physical realizability of predicted flow fields, particularly for extrapolation beyond training data distributions.

5.3 Reduced-Order Models

Reduced-order models (ROMs) aim to capture the essential dynamics of complex physical systems while drastically reducing computational costs compared to full-order models (FOMs). Traditional ROM techniques such as proper orthogonal decomposition (POD), Galerkin projection, and balanced truncation have been successfully applied to a wide range of engineering problems. However, these linear methods often have difficulties with highly nonlinear, parametric, and time-dependent systems. The integration of machine learning with traditional ROM techniques has emerged as a promising avenue to overcome these limitations, leading to more accurate, efficient, and versatile reduced-order models.

5.3.1 Traditional Reduced-Order Modeling Approaches

Proper orthogonal decomposition (POD), also known as Karhunen-Loève decomposition or principal component analysis (PCA) in the discrete setting, represents the cornerstone of traditional ROM techniques. POD identifies an optimal low-dimensional basis from high-fidelity simulation snapshots by solving an eigenvalue problem that maximizes the variance captured by the basis functions. The reduced-order model is then constructed by Galerkin projection of the governing equations onto this POD basis, yielding a system of ordinary differential equations (ODEs) with significantly fewer degrees of freedom than the original discretized partial differential equations.

For parametric systems, POD-Galerkin methods face several challenges. The POD basis is typically constructed for a specific parameter configuration, and its accuracy deteriorates for different parameter values. This limitation necessitates either global POD bases that may be unnecessarily large or local POD bases with interpolation strategies. Furthermore, the computational cost of evaluating nonlinear terms in the projected equations can limit the overall speedup, particularly when dealing with state dependent nonlinearities. For systems with moving shocks, discontinuities, or sharp gradients, linear POD may require an impractically large number of modes to achieve acceptable accuracy.

5.3.2 Machine Learning-Enhanced Dimensionality Reduction

Autoencoders, particularly convolutional autoencoders (CAEs), have emerged as powerful tools for nonlinear dimensionality reduction in reduced-order modeling. Unlike linear POD, autoencoders can learn nonlinear manifolds that more efficiently represent complex solution spaces. The encoder network compresses high-dimensional state variables into a low-dimensional latent representation, while the decoder reconstructs the full order solution from this compact representation. This nonlinear compression often achieves superior accuracy compared to POD with fewer latent dimensions [79, 80].

Convolutional autoencoders leverage spatial structure in the data, making them particularly effective for problems involving spatially-distributed fields such as fluid flows, structural deformations, and temperature distributions. Multiple recent studies have demonstrated that CAEs can achieve reconstruction errors several orders of magnitude lower than POD for the same latent dimension, especially for problems with complex spatial patterns or discontinuities [81, 82]. Variational autoencoders (VAEs) extend this capability by learning probabilistic latent representations, enabling uncertainty quantification and generation of new samples within the learned manifold [83].

For time-dependent problems, temporal convolutional autoencoders provide an additional layer of compression by identifying coherent temporal patterns. This approach enables simultaneous spatial and temporal reduction, where one CAE compresses the spatial dimension and another handles the temporal evolution [82]. The resulting ultra-compact representation facilitates efficient storage, transmission, and analysis of large-scale spatiotemporal datasets from high-fidelity simulations.

5.3.3 Hybrid POD-Neural Network Approaches

Combining the interpretability and theoretical foundations of POD with the flexibility of neural networks has led to several successful hybrid approaches. In the POD-DL-ROM framework, POD provides an initial dimensionality reduction, and deep neural networks learn the temporal evolution in the reduced space or approximate nonlinear correction terms [84]. This combination leverages POD's ability to extract physically meaningful modes while using neural networks to capture complex nonlinear dynamics that are difficult to model with traditional Galerkin projection.

POD-Galerkin models augmented with neural networks address the Kolmogorov barrier to reducibility, which states that linear methods may require exponentially many modes for certain classes of problems. Neural networks can approximate unresolved dynamics or closure terms that account for the truncated POD modes, significantly improving ROM accuracy without increasing the latent dimension [85]. For flow in porous media, neural networks have been employed to model projected Forchheimer terms and capture the influence of unresolved modes, enabling accurate predictions with minimal computational cost [86].

The Galerkin neural network-POD method combines POD basis construction with neural network-based approximation of solution coefficients. This approach has demonstrated excellent performance for parametric wave propagation problems, enabling rapid evaluation of reduced-order solutions across the parameter space [87]. The neural network learns a mapping from parameters to POD coefficients, effectively creating a parametric ROM that can be evaluated in real-time.

5.3.4 Neural Network Architectures for Time Evolution

Long short-term memory (LSTM) networks and gated recurrent units (GRUs) naturally handle sequential data and have been successfully applied to predict the temporal evolution of reduced-order states. These recurrent architectures can learn complex temporal dependencies and long-term correlations in the dynamics, making them suitable for problems with history-dependent behaviors or chaotic dynamics. Recent work has shown that LSTMs can accurately predict the evolution of POD coefficients for turbulent flows, structural dynamics, and other time-dependent systems [88].

Temporal convolutional networks (TCNs) and transformer architectures offer alternative approaches to sequence modeling, often achieving superior performance compared to RNNs for long sequences. TCNs use dilated causal convolutions to capture long-range temporal dependencies efficiently, while transformers leverage attention mechanisms to learn relevant temporal correlations across arbitrary time scales. These architectures have demonstrated promise for ROM applications requiring accurate long-time predictions or multi-scale temporal dynamics.

Neural ordinary differential equations (Neural ODEs) provide a continuous-time framework for learning dynamical systems. Instead of discretizing time evolution, Neural ODEs parameterize the right-hand side of an ODE system with a neural network, and the solution is computed using adaptive ODE solvers. This approach offers several advantages including memory efficiency during training, continuous-time predictions, and the ability to handle irregularly sampled data. Neural ODEs have been applied to learn reduced-order dynamics from sparse or noisy observations of complex systems.

5.3.5 Generative Models for Reduced-Order Modeling

Generative adversarial networks (GANs) have been adapted for reduced-order modeling through the GAROM (Generative Adversarial Reduced Order Modelling) framework [89]. In GAROM, the generator learns to produce solutions to parametric differential equations conditioned on problem parameters, while the discriminator (structured as an autoencoder) distinguishes between generated solutions and high-fidelity training data. This adversarial training encourages the generator to capture the true distribution of solutions across the parameter space, potentially yielding more robust ROMs compared to deterministic approaches.

The conditioning mechanism in GAROM allows the model to generalize across different parameter configurations, material properties, or boundary conditions. By training on a diverse set of high-fidelity simulations, GAROM can generate accurate approximations for new parameter combinations without requiring additional full-order simulations. This capability is particularly valuable for design optimization, uncertainty quantification, and parametric studies where numerous evaluations are needed across a wide parameter range.

Diffusion models and normalizing flows represent alternative generative modeling approaches that have recently gained attention in scientific computing. These models can learn complex probability distributions over solution spaces and enable sampling of multiple plausible solutions for problems with inherent uncertainty or multiple equilibria. While their application to ROM is still emerging, preliminary results suggest they may offer advantages in terms of training stability and sample quality compared to GANs.

5.3.6 Applications to Digital Twins and Real-Time Simulation

Reduced-order models enhanced by machine learning serve as critical components in digital twin frameworks, which require real-time or near-real-time simulation capabilities [90]. A digital twin combines sensor data, reduced-order models, and machine learning to create a continuously updated virtual representation of a physical system. ML-enhanced ROMs enable the rapid evaluation of different scenarios, predictive maintenance scheduling, and optimal control strategies.

For manufacturing applications, ROMs trained on comprehensive finite element analyses can predict stress distributions, temperature fields, and deformation patterns in real-time as production parameters change. Machine learning models predict optimal ROM configurations based on current operating conditions, achieving speedups of several orders of magnitude compared to full-order simulations [91]. This capability enables closed-loop optimization where the digital twin guides manufacturing processes to achieve desired outcomes while adapting to changing conditions.

In thermal management systems, predictive digital twins leveraging ML-enhanced ROMs can anticipate thermal behavior under varying loads and environmental conditions [92]. The combination of reduced-order physics models and data-driven correction terms allows these systems to maintain accuracy even when physical models are imperfect or uncertain. Interpretable machine learning techniques can identify key features and dominant modes, providing insights into system behavior that inform design improvements.

5.3.7 Perspectives on ML-ROM Integration

Several research directions show particular promise for advancing the integration of machine learning with reduced-order modeling. Physics-informed architectures that embed conservation laws, symmetries, and other physical constraints directly into neural network structures can improve generalization, reduce data requirements, and ensure physical realizability of predictions. Structure-preserving neural networks that respect symplectic geometry, energy conservation, or other invariants represent an important frontier for ROM applications in conservative systems.

Transfer learning and domain adaptation techniques can enable ROMs trained on one system or parameter regime to be efficiently adapted to related problems with minimal additional data. This capability is particularly valuable for problems where high-fidelity simulations are extremely expensive or where only limited data is available for the target configuration. Multi-fidelity approaches that combine information from simulations at different resolutions or from different physical models offer another avenue to enhance ROM accuracy and efficiency.

Uncertainty quantification remains a critical challenge for ML-enhanced ROMs, particularly for safety-critical applications. Bayesian neural networks, ensemble methods, and conformal prediction provide frameworks for quantifying epistemic and aleatoric uncertainty in ROM predictions. Understanding when and how ROMs fail, and providing reliable uncertainty estimates, will be essential for deploying these methods in engineering practice. Recent work on adversarial robustness and out-of-distribution detection also has important implications for ensuring ROM reliability.

The development of interpretable and explainable ML-ROM frameworks will facilitate adoption by practitioners and enable new scientific insights. Attention mechanisms can identify which spatial regions or temporal intervals are most important for predictions. Symbolic regression and equation discovery techniques can extract governing equations or closure models from learned neural network dynamics, potentially revealing new physics or validating modeling assumptions.

Scalability to very high-dimensional systems and real-time constraints poses ongoing challenges. Graph neural networks and mesh-free representations offer potential solutions for handling irregular geometries and adaptive refinement. Distributed and federated learning approaches can enable ROM training on massive datasets distributed across multiple sites or organizations. Hardware acceleration through GPUs, TPUs, and specialized AI chips will be increasingly important for achieving the performance required for digital twin applications.

5.4 Digital Twins

A digital twin is a virtual representation of a physical system that evolves in synchronization with its real-world counterpart through continuous data exchange and model updating. Digital twins integrate multi-physics simulations, sensor data, machine learning models, and real-time monitoring to enable predictive analysis, optimization, and decision-making throughout the lifecycle of physical assets. In the context of computational mechanics, digital twins extend beyond static simulation models by incorporating data assimilation, inverse problems, reduced-order modeling, and adaptive learning to maintain fidelity with evolving physical systems. The integration of machine learning has become essential for digital twins, enabling real-time predictions, uncertainty quantification, and automated model adaptation in response to changing conditions.

5.4.1 Digital Twin Architectures and Components

A comprehensive digital twin architecture comprises five interconnected layers: the physical layer (sensors, actuators, and physical assets), the data acquisition layer (communication protocols and edge computing), the digital model layer (physics-based simulations and reduced-order models), the analytics layer (machine learning algorithms and optimization routines), and the visualization and decision-making layer (user interfaces and control systems). The bidirectional data flow between physical and digital domains distinguishes digital twins from conventional simulation models, enabling closed-loop feedback where predictions inform physical operations and sensor measurements continuously update digital models.

The digital model layer typically employs hierarchical modeling strategies, combining high-fidelity finite element models for detailed analysis with reduced-order models for real-time simulation. Machine learning enhances this layer by providing surrogate models that approximate expensive computations, enabling rapid scenario evaluation and parameter studies [91]. Physics-informed neural networks serve as hybrid models that respect governing equations while learning corrections from data, ensuring both computational efficiency and physical consistency [93].

Data assimilation techniques merge sensor observations with model predictions, accounting for uncertainties in both measurements and simulations. Ensemble Kalman filters, particle filters, and variational methods provide frameworks for sequential state estimation, updating model states as new data arrives. Machine learning-based data assimilation methods, including neural network state estimators and deep Kalman filters, offer advantages for high-dimensional nonlinear systems where traditional methods become computationally prohibitive.

5.4.2 Model Updating and Inverse Problems

Digital twins require continuous model updating to maintain alignment with physical systems as they degrade, experience damage, or operate under varying conditions. Traditional model updating approaches based on optimization and Bayesian inference can be computationally expensive, limiting their applicability for real-time digital twins. The inverse mapping parameter updating (IMPU) method provides an efficient alternative by training neural networks to directly map sensor measurements to physically interpretable model parameters [94].

Bayesian approaches to inverse problems enable rigorous uncertainty quantification in parameter estimates, providing probability distributions rather than point estimates. Recent work has extended IMPU with Bayesian neural networks to infer parameter distributions in real-time, enabling probabilistic predictions that account for epistemic uncertainty [95]. This capability is crucial for safety-critical applications where decision-making must consider the confidence level of digital twin predictions.

For structural systems, model updating typically involves identifying changes in stiffness, damping, or boundary conditions from vibration measurements or strain gauge data. Machine learning methods can extract damage-sensitive features from high-dimensional sensor data, reducing the inverse problem dimension and improving identifiability. Convolutional neural networks process spatial sensor arrays, while recurrent networks capture temporal dependencies in dynamic response data. Physics-decoded variational neural networks integrate reduced-order digital twins as decoders within variational frameworks, seamlessly combining data-driven learning with physics-based constraints [96].

5.4.3 Predictive Maintenance and Structural Health Monitoring

Predictive maintenance represents one of the most impactful applications of digital twins in industrial settings. By combining real-time sensor data with predictive models, digital twins forecast equipment failures before they occur, enabling scheduled maintenance that minimizes downtime and reduces costs. Machine learning algorithms analyze patterns in vibration, temperature, acoustic emission, and other sensor modalities to detect anomalies and predict remaining useful life [97].

Deep learning architectures have demonstrated significant advantages for predictive maintenance tasks. Autoencoders learn compact representations of normal operating conditions and flag anomalies when reconstruction errors exceed thresholds [98]. Recurrent neural networks, particularly LSTMs, capture temporal degradation patterns and forecast future health states based on historical trends. Recent implementations have achieved significant improvements in predictive accuracy and reductions in unplanned downtime compared to traditional rule-based maintenance strategies [99].

For structural health monitoring, digital twins integrate sensor networks with finite element models to continuously assess structural integrity. Graph neural networks provide natural architectures for representing structural systems, with nodes corresponding to structural members and edges encoding connectivity. The G-Twin framework applies graph neural networks to offshore wind turbine monitoring, achieving real-time high-fidelity predictions of structural response under varying environmental conditions [100]. Transfer learning enables digital twins trained on one structure to be adapted to similar structures with minimal additional data, reducing the cost of deployment across asset portfolios.

Hybrid digital twins combine physics-based models with data-driven corrections, leveraging the complementary strengths of both approaches. Physics-informed neural networks serve as surrogate models within digital twin frameworks, embedding governing equations while learning from data to correct model discrepancies [101, 102]. This hybrid approach maintains interpretability and extrapolation capabilities beyond purely data-driven methods while achieving computational efficiency required for real-time applications.

5.4.4 Real-Time Simulation and Cyber-Physical Systems

Real-time simulation capabilities are fundamental to digital twin functionality, enabling continuous synchronization with physical systems and rapid evaluation of what-if scenarios. Achieving real-time performance for complex computational mechanics problems requires aggressive model reduction, often by factors of 3–6 orders of magnitude. Machine learning-enhanced reduced-order models provide the necessary computational speedup while maintaining acceptable accuracy [103].

Cyber-physical systems (CPS) integrate computation, networking, and physical processes, with digital twins serving as the computational core that mirrors physical behavior. The bidirectional coupling between digital and physical domains enables closed-loop control where digital twin predictions inform actuator commands, and sensor feedback continuously refines digital models. This tight coupling raises challenges in synchronization, latency management, and ensuring stability of the coupled system.

Edge computing architectures distribute computation between local devices and the cloud, balancing latency requirements with computational resources. Smaller neural networks deployed on edge devices perform inference in real-time, while more complex model training and updating occurs in the cloud. Federated learning frameworks enable collaborative digital twin development across multiple assets while preserving data privacy and reducing communication overhead.

For manufacturing systems, digital twins enable real-time process monitoring, quality control, and optimization. Sensor data from production lines feeds into digital models that predict product quality, identify defects, and recommend process adjustments. Reinforcement learning agents trained in digital twin environments can discover optimal control policies before deployment on physical systems, reducing trial-and-error experimentation and accelerating process improvement [104].

5.4.5 Uncertainty Quantification and Decision-Making

Uncertainty quantification (UQ) is essential for digital twins supporting critical decisions, as predictions must be accompanied by confidence bounds that reflect model uncertainties, measurement noise, and inherent system variability. Bayesian frameworks provide principled approaches to UQ, propagating uncertainties from inputs through models to outputs. However, traditional UQ methods face computational challenges for complex nonlinear systems, motivating the development of machine learning-enhanced UQ techniques.

Bayesian neural networks and neural network ensembles provide probabilistic predictions with uncertainty estimates. Dropout-based uncertainty quantification offers a computationally efficient approximation to Bayesian inference, enabling uncertainty estimation without explicit probabilistic modeling. Conformal prediction provides distribution-free uncertainty quantification with finite-sample validity guarantees, creating prediction intervals with specified coverage probabilities.

Decision-making under uncertainty requires trading off multiple objectives while accounting for incomplete knowledge. Digital twins support decision analysis by rapidly evaluating candidate actions across uncertain scenarios, computing expected outcomes and risk metrics. Reinforcement learning combined with digital twin simulation enables learning of decision policies that optimize long-term objectives while managing uncertainty. Explainable AI techniques help stakeholders understand the basis for digital twin recommendations, building trust and facilitating informed decision-making.

5.4.6 *Perspectives on Digital Twin Development*

The future development of digital twins will be shaped by several key research directions and technological trends. Standardization of digital twin architectures, data formats, and interfaces will facilitate interoperability and enable ecosystem development around digital twin platforms. Open-source frameworks for digital twin development can accelerate adoption and foster collaboration between academia and industry.

Autonomous digital twins that self-adapt to changing conditions represent an important frontier. Online learning algorithms continuously update models as new data arrives, detecting concept drift and triggering model retraining when performance degrades. Meta-learning approaches enable rapid adaptation to new scenarios by learning algorithms optimized for quick fine-tuning. Automated machine learning (AutoML, see Sect. 2.5.1) can optimize neural network architectures and hyperparameters specifically for each digital twin application.

Federated digital twin networks where multiple digital twins collaborate and share knowledge offer potential for improved performance and robustness. Transfer learning enables knowledge gained from one digital twin to inform others representing similar systems. Distributed optimization allows coordinated decision-making across interconnected assets, such as building climate control systems or power grid operations.

As computational capabilities continue advancing and sensor technologies become more ubiquitous, digital twins will extend to larger-scale systems including smart cities, infrastructure networks, and environmental systems. Climate digital twins aim to predict regional climate evolution, inform adaptation strategies, and evaluate intervention effectiveness. Challenges include handling massive spatiotemporal datasets, quantifying uncertainties in chaotic systems, and integrating human behavioral models.

References

1. Li, X., Liu, Z., Cui, S., Luo, C., Li, C., Zhuang, Z.: Predicting the effective mechanical property of heterogeneous materials by image based modeling and deep learning. Comput. Methods Appl. Mech. Eng. **347**, 735–753 (2019). 10.1016/j.cma.2019.01.005
2. Minh Nguyen-Thanh, V., Trong Khiem Nguyen, L., Rabczuk, T., Zhuang, X.: A surrogate model for computational homogenization of elastostatics at finite strain using high-dimensional model representation-based neural network. Int. J. Numer. Methods Eng. **121**(21), 4811–4842 (2020). 10.1002/nme.6493
3. Linka, K., Hillgärtner, M., Abdolazizi, K.P., Aydin, R.C., Itskov, M., Cyron, C.J.: Constitutive artificial neural networks: a fast and general approach to predictive data-driven constitutive modeling by deep learning. J. Comput. Phys. **429**, 110010 (2021). 10.1016/j.jcp.2020.110010
4. Xu, R., Yang, J., Yan, W., Huang, Q., Giunta, G., Belouettar, S., et al.: Data-driven multiscale finite element method: From concurrence to separation. Comput. Methods Appl. Mech. Eng. **363**, 112893 (2020). 10.1016/j.cma.2020.112893

5. Yan, S., Zou, X., Ilkhani, M., Jones, A.: An efficient multiscale surrogate modelling framework for composite materials considering progressive damage based on artificial neural networks. Compos. Part B Eng. **194**, 108014 (2020). 10.1016/j.compositesb.2020.108014
6. Liu, Z., Wu, C., Koishi, M.: A deep material network for multiscale topology learning and accelerated nonlinear modeling of heterogeneous materials. Comput. Methods Appl. Mech. Eng. **345**, 1138–1168 (2019). 10.1016/j.cma.2018.09.020
7. Liu, Z., Wu, C.: Exploring the 3D architectures of deep material network in data-driven multiscale mechanics. J. Mech. Phys. Solids **127**, 20–46 (2019). 10.1016/j.jmps.2019.03.004
8. Mozaffar, M., Bostanabad, R., Chen, W., Ehmann, K., Cao, J., Bessa, M.: Deep learning predicts path-dependent plasticity. Proc. Natl. Acad. Sci. **116**(52), 26414–26420 (2019). 10.1073/pnas.1911815116
9. Abueidda, D.W., Koric, S., Sobh, N.A., Sehitoglu, H.: Deep learning for plasticity and thermo-viscoplasticity. Int. J. Plast. **136**, 102852 (2021). 10.1016/j.ijplas.2020.102852
10. Ghavamian, F., Simone, A.: Accelerating multiscale finite element simulations of history-dependent materials using a recurrent neural network. Comput. Methods Appl. Mech. Eng. **357**, 112594 (2019). 10.1016/j.cma.2019.112594
11. Wu, L., Kilingar, N.G., Noels, L., et al.: A recurrent neural network-accelerated multi-scale model for elasto-plastic heterogeneous materials subjected to random cyclic and non-proportional loading paths. Comput. Methods Appl. Mech. Eng. **369**, 113234 (2020). 10.1016/j.cma.2020.113234
12. Wang, K., Sun, W.: A multiscale multi-permeability poroplasticity model linked by recursive homogenizations and deep learning. Comput. Methods Appl. Mech. Eng. **334**, 337–380 (2018). 10.1016/j.cma.2018.01.036
13. Yang, Z., Yabansu, Y.C., Al-Bahrani, R., Liao, W.-k., Choudhary, A.N., Kalidindi, S.R., et al.: Deep learning approaches for mining structure-property linkages in high contrast composites from simulation datasets. Comput. Mater. Sci. **151**, 278–287 (2018). 10.1016/j.commatsci.2018.05.014
14. Wei, A., Xiong, J., Yang, W., Guo, F.: Deep learning-assisted elastic isotropy identification for architected materials. Extreme Mech. Lett. **43**, 101173 (2021). 10.1016/j.eml.2021.101173
15. Vasilyeva, M., Tyrylgin, A.: Machine learning for accelerating macroscopic parameters prediction for poroelasticity problem in stochastic media. Comput. Math. Appl. **84**, 185–202 (2021). 10.1016/j.camwa.2020.09.024
16. Rao, C., Liu, Y.: Three-dimensional convolutional neural network (3D-CNN) for heterogeneous material homogenization. Comput. Mater. Sci. **184**, 109850 (2020). 10.1016/j.commatsci.2020.109850
17. Samaei, A., Chaudhuri, S.: Mechanical performance of zirconia-silica bilayer coating on aluminum alloys with varying porosities: Deep learning and microstructure-based FEM. Mater. Des. **207**, 109860 (2021). 10.1016/j.matdes.2021.109860
18. Xiao, S., Hu, R., Li, Z., Attarian, S., Björk, K.-M., Lendasse, A.: A machine-learning-enhanced hierarchical multiscale method for bridging from molecular dynamics to continua. Neural Comput. Appl. **32**(18), 14359–14373 (2020). 10.1007/s00521-019-04480-7
19. Frankel, A.L., Jones, R.E., Alleman, C., Templeton, J.A.: Predicting the mechanical response of oligocrystals with deep learning. Comput. Mater. Sci. **169**, 109099 (2019). 10.1016/j.commatsci.2019.109099
20. Rocha, I., Kerfriden, P., van der Meer, F.: On-the-fly construction of surrogate constitutive models for concurrent multiscale mechanical analysis through probabilistic machine learning. J. Comput. Phys. X **9**, 100083 (2021). 10.1016/j.jcpx.2020.100083
21. Tan, R.K., Zhang, N.L., Ye, W.: A deep learning-based method for the design of microstructural materials. Struct. Multidiscip. Optim. **61**(4), 1417–1438 (2020). 10.1007/s00158-019-02424-2
22. Li, J., Zhang, M., Martins, J.R., Shu, C.: Efficient aerodynamic shape optimization with deep-learning-based geometric filtering. AIAA J. **58**(10), 4243–4259 (2020). 10.2514/1.J059254

23. Yoo, S., Lee, S., Kim, S., Hwang, K.H., Park, J.H., Kang, N.: Integrating deep learning into CAD/CAE system: generative design and evaluation of 3D conceptual wheel. Struct. Multidiscip. Optim. **64**(4), 2725–2747 (2021). 10.1007/s00158-021-02953-9
24. Messner, M.C.: Convolutional neural network surrogate models for the mechanical properties of periodic structures. J. Mech. Des. **142**(2), 024503 (2020). 10.1115/1.4045040
25. Abueidda, D.W., Koric, S., Sobh, N.A.: Topology optimization of 2D structures with nonlinearities using deep learning. Comput. Struct. **237**, 106283 (2020). 10.1016/j.compstruc.2020.106283
26. Kollmann, H.T., Abueidda, D.W., Koric, S., Guleryuz, E., Sobh, N.A.: Deep learning for topology optimization of 2D metamaterials. Mater. Des. **196**, 109098 (2020). 10.1016/j.matdes.2020.109098
27. Bielecki, D., Patel, D., Rai, R., Dargush, G.F.: Multi-stage deep neural network accelerated topology optimization. Struct. Multidiscip. Optim. **64**(6), 3473–3487 (2021). 10.1007/s00158-021-03028-5
28. Chi, H., Zhang, Y., Tang, T.L.E., Mirabella, L., Dalloro, L., Song, L., et al.: Universal machine learning for topology optimization. Comput. Methods Appl. Mech. Eng. **375**, 112739 (2021). 10.1016/j.cma.2019.112739
29. Bi, S., Zhang, J., Zhang, G.: Scalable deep-learning-accelerated topology optimization for additively manufactured materials. Preprint. arXiv:2011.14177 (2020). 10.48550/arXiv.2011.14177
30. Kallioras, N.A., Kazakis, G., Lagaros, N.D.: Accelerated topology optimization by means of deep learning. Struct. Multidiscip. Optim. **62**(3), 1185–1212 (2020). 10.1007/s00158-020-02545-z
31. Deng, H., To, A.C.: Topology optimization based on deep representation learning (DRL) for compliance and stress-constrained design. Comput. Mech. **66**(2), 449–469 (2020). 10.1007/s00466-020-01859-5
32. Chandrasekhar, A., Suresh, K.: TOuNN: topology optimization using neural networks. Struct. Multidiscip. Optim. **63**(3), 1135–1149 (2021). 10.1007/s00158-020-02748-4
33. Wang, T., Altabey, W.A., Noori, M., Ghiasi, R.: A deep learning based approach for response prediction of beam-like structures. Struct. Durability Health Monit. **14**(4), 315 (2020). 10.32604/sdhm.2020.011083
34. Trivedi, R., Su, L., Lu, J., Schubert, M.F., Vuckovic, J.: Data driven acceleration of photonic simulations. Sci. Rep. **9**(1), 1–7 (2019). 10.1038/s41598-019-56212-5
35. Nikolopoulos, S., Kalogeris, I., Papadopoulos, V.: Non-intrusive surrogate modeling for parametrized time-dependent partial differential equations using convolutional autoencoders. Eng. Appl. Artif. Intell. **109**, 104652 (2022). 10.1016/j.engappai.2021.104652
36. Jin, W., Peng, S., Tan, S.X.-D.: Data-driven electrostatics analysis based on physics-constrained deep learning. In: 2021 Design, Automation & Test in Europe Conference & Exhibition (DATE), pp. 1382–1387. IEEE (2021). 10.23919/DATE51398.2021.9474231
37. Zhu, Y., Zabaras, N., Koutsourelakis, P.-S., Perdikaris, P.: Physics-constrained deep learning for high-dimensional surrogate modeling and uncertainty quantification without labeled data. J. Comput. Phys. **394**, 56–81 (2019). 10.1016/j.jcp.2019.05.024
38. Sun, L., Gao, H., Pan, S., Wang, J.-X.: Surrogate modeling for fluid flows based on physics-constrained deep learning without simulation data. Comput. Methods Appl. Mech. Eng. **361**, 112732 (2020). 10.1016/j.cma.2019.112732
39. Luo, X., Kareem, A.: Deep convolutional neural networks for uncertainty propagation in random fields. Comput. Aided Civil Infrastruct. Eng. **34**(12), 1043–1054 (2019). 10.1111/mice.12510
40. Liang, L., Liu, M., Martin, C., Sun, W.: A deep learning approach to estimate stress distribution: a fast and accurate surrogate of finite-element analysis. J. R. Soc. Interface **15**(138), 20170844 (2018). 10.1098/rsif.2017.0844
41. Liang, L., Liu, M., Martin, C., Sun, W.: A machine learning approach as a surrogate of finite element analysis-based inverse method to estimate the zero-pressure geometry of human thoracic aorta. Int. J. Numer. Methods Biomed. Eng. **34**(8), e3103 (2018). 10.1002/cnm.3103

42. Misaka, T., Herwan, J., Kano, S., Sawada, H., Furukawa, Y.: Deep neural network-based cost function for metal cutting data assimilation. Int. J. Adv. Manuf. Technol. **107**, 385–398 (2020). 10.1007/s00170-020-04984-w
43. Papadopoulos, V., Soimiris, G., Giovanis, D., Papadrakakis, M.: A neural network-based surrogate model for carbon nanotubes with geometric nonlinearities. Comput. Methods Appl. Mech. Eng. **328**, 411–430 (2018). 10.1016/j.cma.2017.09.010
44. Chen, G., Li, T., Chen, Q., Ren, S., Wang, C., Li, S.: Application of deep learning neural network to identify collision load conditions based on permanent plastic deformation of shell structures. Comput. Mech. **64**, 435–449 (2019). 10.1007/s00466-019-01706-2
45. Wang, Y., Oyen, D., Guo, W., Mehta, A., Scott, C.B., Panda, N., et al.: StressNetDeep learning to predict stress with fracture propagation in brittle materials. NPJ Mater. Degradation **5**(1), 6 (2021). 10.1038/s41529-021-00151-y
46. Ninić, J., Freitag, S., Meschke, G.: A hybrid finite element and surrogate modelling approach for simulation and monitoring supported TBM steering. Tunn. Underground Space Technol. **63**, 12–28 (2017). 10.1016/j.tust.2016.12.004
47. Koeppe, A., Bamer, F., Markert, B.: An efficient Monte Carlo strategy for elasto-plastic structures based on recurrent neural networks. Acta Mech. **230**, 3279–3293 (2019). 10.1007/s00707-019-02436-5
48. Cao, B.-T., Obel, M., Freitag, S., Mark, P., Meschke, G.: Artificial neural network surrogate modelling for real-time predictions and control of building damage during mechanised tunnelling. Adv. Eng. Softw. **149**, 102869 (2020). 10.1016/j.advengsoft.2020.102869
49. Peng, H., Yan, J., Yu, Y., Luo, Y.: Structural surrogate model and dynamic response prediction with consideration of temporal and spatial evolution: An encoder–decoder ConvLSTM network. Int. J. Struct. Stab. Dyn. **21**(10), 2150140 (2021). 10.1142/S0219455421501406
50. Zhang, L., Cheng, L., Li, H., Gao, J., Yu, C., Domel, R., et al.: Hierarchical deep-learning neural networks: finite elements and beyond. Comput. Mech. **67**, 207–230 (2021). 10.1007/s00466-020-01928-9
51. Oishi, A., Yagawa, G.: Computational mechanics enhanced by deep learning. Comput. Methods Appl. Mech. Eng. **327**, 327–351 (2017). 10.1016/j.cma.2017.08.040
52. Koeppe, A., Bamer, F., Markert, B.: An intelligent nonlinear meta element for elastoplastic continua: deep learning using a new Time-distributed Residual U-Net architecture. Comput. Methods Appl. Mech. Eng. **366**, 113088 (2020). 10.1016/j.cma.2020.113088
53. Bao, H., Feng, J., Dinh, N., Zhang, H.: Computationally efficient CFD prediction of bubbly flow using physics-guided deep learning. Int. J. Multiphase Flow **131**, 103378 (2020). 10.1016/j.ijmultiphaseflow.2020.103378
54. Wang, Y., Lin, G.: Efficient deep learning techniques for multiphase flow simulation in heterogeneous porousc media. J. Comput. Phys. **401**, 108968 (2020). 10.1016/j.jcp.2019.108968
55. Halder, R., Damodaran, M., Khoo, B.: Deep learning based reduced order model for airfoil-gust and aeroelastic interaction. AIAA J. **58**(10), 4304–4321 (2020). 10.2514/1.J059027
56. Srivastava, S., Damodaran, M., Khoo, B.C.: Machine learning surrogates for predicting response of an aero-structural-sloshing system. Preprint. arXiv:1911.10043 (2019). 10.48550/arXiv.1911.10043
57. Geneva, N., Zabaras, N.: Multi-fidelity generative deep learning turbulent flows. Preprint. arXiv:2006.04731 (2020). 10.3934/fods.2020019
58. Obiols-Sales, O., Vishnu, A., Malaya, N., Chandramowliswharan, A.: CFDNet: A deep learning-based accelerator for fluid simulations. In: Proceedings of the 34th ACM International Conference on Supercomputing, pp. 1–12 (2020). 10.1145/3392717.3392772
59. Guo, X., Li, W., Iorio, F.: Convolutional neural networks for steady flow approximation. In: Proceedings of the 22nd ACM SIGKDD International Conference on Knowledge Discovery and Data Mining, pp. 481–490 (2016). 10.1145/2939672.2939738
60. Takbiri-Borujeni, A., Kazemi, H., Nasrabadi, N.: A data-driven surrogate to image-based flow simulations in porous media. Comput. Fluids **201**, 104475 (2020). 10.1016/j.compfluid.2020.104475

61. Tang, M., Liu, Y., Durlofsky, L.J.: A deep-learning-based surrogate model for data assimilation in dynamic subsurface flow problems. J. Comput. Phys. **413**, 109456 (2020). 10.1016/j.jcp.2020.109456
62. Tang, M., Liu, Y., Durlofsky, L.J.: Deep-learning-based surrogate flow modeling and geological parameterization for data assimilation in 3D subsurface flow. Comput. Methods Appl. Mech. Eng. **376**, 113636 (2021). 10.1016/j.cma.2020.113636
63. Tang, M., Ju, X., Durlofsky, L.J.: Deep-learning-based coupled flow-geomechanics surrogate model for CO2 sequestration. Int. J. Green-house Gas Control **118**, 103692 (2022). 10.1016/j.ijggc.2022.103692
64. Rana, P., Weigand, T.M., Pilkiewicz, K.R., Mayo, M.L.: A scalable convolutional neural network approach to fluid flow prediction in complex environments. Sci. Rep. **14**(1), 23080 (2024). 10.1038/s41598-024-73529-y
65. Fowler, E., McDevitt, C.J., Roy, S.: Physics-informed neural network simulation of thermal cavity flow. Sci. Rep. **14**(1), 15203 (2024). 10.1038/s41598-024-65664-3
66. Pioch, F., Harmening, J.H., Müller, A.M., Peitzmann, F.-J., Schramm, D., Moctar, O.e.: Turbulence modeling for physics-informed neural networks: Comparison of different RANS models for the backward-facing step flow. Fluids **8**(2), 43 (2023). 10.3390/fluids8020043
67. Zhang, J., Braga-Neto, U., Gildin, E.: Physics-informed neural networks for multiphase flow in porous media considering dual shocks and interphase solubility. Energy Fuels **38**(18), 17781–17795 (2024). 10.1021/acs.energyfuels.4c02888
68. Qi, X., de Almeida, G.A., Maldonado, S.: Physics-informed neural networks for solving flow problems modeled by the 2D Shallow Water Equations without labeled data. J. Hydrol. **636**, 131263 (2024). 10.1016/j.jhydrol.2024.131263
69. Zhou, W., Miwa, S., Okamoto, K.: Advancing fluid dynamics simulations: A comprehensive approach to optimizing physics-informed neural networks. Phys. Fluids **36**(1) (2024). 10.1063/5.0180770
70. Buaria, D., Sreenivasan, K.R.: Forecasting small-scale dynamics of fluid turbulence using deep neural networks. Proc. Natl. Acad. Sci. **120**(30), e2305765120 (2023). 10.1073/pnas.2305765120
71. Yousif, M.Z., Yu, L., Hoyas, S., Vinuesa, R., Lim, H.: A deep-learning approach for reconstructing 3D turbulent flows from 2D observation data. Sci. Rep. **13**(1), 2529 (2023). 10.1038/s41598-023-29525-9
72. Lei, B., Fu, Y., Cadena, J., Saini, A., Hu, Y., Bao, J., et al.: Accelerating computational fluid dynamics simulation of post-combustion carbon capture modeling with MeshGraphNets. Front. Artif. Intell. **7**, 1441985 (2025). 10.3389/frai.2024.1441985
73. Li, Z., Huang, D.Z., Liu, B., Anandkumar, A.: Fourier neural operator with learned deformations for PDEs on general geometries. J. Mach. Learn. Res. **24**(388), 1–26 (2023). http://jmlr.org/papers/v24/23-0064.html
74. Font, B., Alcántara-Ávila, F., Rabault, J., Vinuesa, R., Lehmkuhl, O.: Deep reinforcement learning for active flow control in a turbulent separation bubble. Nat. Commun. **16**(1), 1422 (2025). 10.1038/s41467-025-56408-6
75. Tao, J., Sun, G.: Application of deep learning based multi-fidelity surrogate model to robust aerodynamic design optimization. Aerospace Sci. Technol. **92**, 722–737 (2019). 10.1016/j.ast.2019.07.002
76. Cai, S., Mao, Z., Wang, Z., Yin, M., Karniadakis, G.E.: Physics-informed neural networks (PINNs) for fluid mechanics: a review. Acta Mech. Sin. **37**(12), 1727–1738 (2021). 10.1007/s10409-021-01148-1
77. Ang, E.H.W., Wang, G., Ng, B.F.: Physics-informed neural networks for low Reynolds number flows over cylinder. Energies **16**(12), 4558 (2023). 10.3390/en16124558
78. Pfaff, T., Fortunato, M., Sanchez-Gonzalez, A., Battaglia, P.W.: Learning mesh-based simulation with graph networks. Preprint. arXiv:2010.03409 (2021). 10.48550/arXiv.2010.03409
79. Fresca, S., Dede', L., Manzoni, A.: A comprehensive deep learning-based approach to reduced order modeling of nonlinear time-dependent parametrized PDEs. J. Sci. Comput. **87**(2), 61 (2021). 10.1007/s10915-021-01462-7

80. Lee, K., Carlberg, K.T.: Model reduction of dynamical systems on nonlinear manifolds using deep convolutional autoencoders. J. Comput. Phys. **404**, 108973 (2020). 10.1016/j.jcp.2019.108973
81. Venkat, S., Smith, R.C., Kelley, C.T.: Convolutional autoencoders for reduced-order modeling. Preprint. arXiv:2108.12453 (2021). 10.48550/arXiv.2108.12453
82. Abdedou, A., Soulaimani, A.: Reduced-order modeling for stochastic large-scale and time-dependent flow problems using deep spatial and temporal convolutional autoencoders. Adv. Model. Simul. Eng. Sci. **10**(1), 7 (2023). 10.1186/s40323-023-00244-0
83. Simpson, T., Vlachas, K., Garland, A., Dervilis, N., Chatzi, E.: VpROM: a novel variational autoencoder-boosted reduced order model for the treatment of parametric dependencies in nonlinear systems. Sci. Rep. **14**(1), 6091 (2024). 10.1038/s41598-024-56118-x
84. Fresca, S., Manzoni, A.: POD-DL-ROM: Enhancing deep learning-based reduced order models for nonlinear parametrized PDEs by proper orthogonal decomposition. Comput. Methods Appl. Mech. Eng. **388**, 114181 (2021). 10.1016/j.cma.2021.114181
85. Barnett, J., Farhat, C., Maday, Y.: Neural network augmented projection-based model order reduction for mitigating the Kolmogorov barrier to reducibility. J. Comput. Phys. **492**, 112420 (2023). 10.1016/j.jcp.2023.112420
86. Allery, C., Béghein, C., Dubot, C., Dubot, F.: POD-Galerkin reduced order model coupled with neural networks to solve flow in porous media. J. Comput. Sci. **84**, 102471 (2025). 10.1016/j.jocs.2024.102471
87. Weder, P., Kast, M., Henríquez, F., Hesthaven, J.S.: Galerkin neural network-POD for acoustic and electromagnetic wave propagation in parametric domains. Adv. Comput. Math. **51**(4), 37 (2025). 10.1007/s10444-025-10245-9
88. Hasegawa, K., Fukami, K., Murata, T., Fukagata, K.: Machine-learning-based reduced-order modeling for unsteady flows around bluff bodies of various shapes. Theoret. Comput. Fluid Dyn. **34**(4), 367–383 (2020). 10.1007/s00162-020-00528-w
89. Coscia, D., Demo, N., Rozza, G.: Generative adversarial reduced order modelling. Sci. Rep. **14**(1), 3590 (2024). 10.1038/s41598-024-54067-z
90. Ghnatios, C., Rodriguez, S., Tomezyk, J., Dupuis, Y., Mouterde, J., Da Silva, J., et al.: A hybrid twin based on machine learning enhanced reduced order model for real-time simulation of magnetic bearings. Adv. Model. Simul. Eng. Sci. **11**(1), 3 (2024). 10.1186/s40323-024-00258-2
91. Kapteyn, M.G., Knezevic, D.J., Willcox, K.: Toward predictive digital twins via component-based reduced-order models and interpretable machine learning. In: AIAA scitech 2020 Forum, p. 0418 (2020). 10.2514/6.2020-0418
92. Subramani, T., Bartscher, S.: Predictive digital twins for thermal management using machine learning and reduced-order models. Preprint. arXiv:2505.06849 (2025). 10.48550/arXiv.2505.06849
93. Willard, J., Jia, X., Xu, S., Steinbach, M., Kumar, V.: Integrating scientific knowledge with machine learning for engineering and environmental systems. ACM Comput. Surv. **55**(4), 1–37 (2022). 10.1145/3514228
94. Kessels, B.M., Fey, R.H., van de Wouw, N.: Real-time parameter updating for nonlinear digital twins using inverse mapping models and transient-based features. Nonlinear Dyn. **111**(11), 10255–10285 (2023). 10.1007/s11071-023-08354-5
95. Kessels, B.M., Fey, R.H., van de Wouw, N.: Uncertainty quantification in real-time parameter updating for digital twins using Bayesian inverse mapping models. Nonlinear Dyn. **113**(8), 7613–7637 (2025). 10.1007/s11071-024-10608-9
96. Xu, J., Zhou, X., Giglio, M., Sbarufatti, C., Dong, L.: Variational neural network embedded with digital twins for probabilistic structural damage quantification. AIAA J. **63**(6), 2474–2486 (2025). 10.2514/1.J064808
97. Fuller, A., Fan, Z., Day, C., Barlow, C.: Digital twin: Enabling technologies, challenges and open research. IEEE Access **8**, 108952–108971 (2020). 10.1109/ACCESS.2020.2998358

98. Pota, M., De Pietro, G., Esposito, M.: Real-time anomaly detection on time series of industrial furnaces: A comparison of autoencoder architectures. Eng. Appl. Artif. Intell. **124**, 106597 (2023). 10.1016/j.engappai.2023.106597
99. Theissler, A., Pé rez-Velázquez, J., Kettelgerdes, M., Elger, G.: Predictive maintenance enabled by machine learning: use cases and challenges in the automotive industry. Reliab. Eng. Syst. Saf. **215**, 107864 (2021). 10.1016/j.ress.2021.107864
100. Jiang, C., Chen, N.-Z.: G-Twin: Graph neural network-based digital twin for real-time and high-fidelity structural health monitoring for offshore wind turbines. Marine Struct. **103**, 103813 (2025). 10.1016/j.marstruc.2025.103813
101. Chinesta, F., Cueto, E., Abisset-Chavanne, E., Duval, J.L., El Khaldi, F.: Virtual, digital and hybrid twins: a new paradigm in data-based engineering and engineered data. Arch. Comput. Methods Eng. **27**(1), 105–134 (2018). 10.1007/s11831-018-9301-4
102. Sahin, T., Wolff, D., von Danwitz, M., Popp, A.: Towards a hybrid digital twin: Physics-informed neural networks as surrogate model of a reinforced concrete beam. Preprint. arXiv:2405.08406 (2024). 10.48550/arXiv.2405.08406
103. Xu, X., Wang, G., Yan, H., Zhang, L., Yao, X.: Deep-learning-enhanced digital twinning of complex composite structures and real-time mechanical interaction. Compos. Sci. Technol. **241**, 110139 (2023). 10.1016/j.compscitech.2023.110139
104. Melesse, T.Y., Di Pasquale, V., Riemma, S.: Digital twin models in industrial operations: a systematic literature review. Procedia Manuf. **42**, 267–272 (2020). 10.1016/j.promfg.2020.02.084

Chapter 6
Neural Operators

6.1 Introduction of Neural Operator

Neural operators learn mappings between functions, and have found extensive applications in science and engineering [1]. For instance, the input function may represent the boundary conditions of a partial differential equation (PDE), while the output corresponds to the solution field of interest. Although functions are infinite-dimensional in mathematical terms, in practice, we typically input discretized data defined on a finite mesh to approximate continuous functions.

Unlike traditional data-driven methods based on neural networks, operator learning is characterized by *discretization-invariance*, which means:

- The ability to input at any resolution.
- The capability to output at any location.
- Convergence of results as the mesh is refined.

Conventional neural network-based models are often sensitive to the resolution of the input and output; if the discretization changes, the model typically requires retraining. In contrast, operator learning models preserve their performance across different resolutions, owing to their character of discretization-invariant.

Furthermore, not all function-fitting algorithms can be classified as neural operators. To be considered a neural operator, the model must satisfy both discretization-invariance and the generalized approximation property [2]. Neural operators thus represent a superior class of algorithms within the broader framework of operator learning, with notable examples such as the Fourier Neural Operator (FNO) [3].

T. Rabczuk et al., *Scientific Machine Learning with Engineering Applications*,
Studies in Systems, Decision and Control 667,
https://doi.org/10.1007/978-3-032-20307-6_6

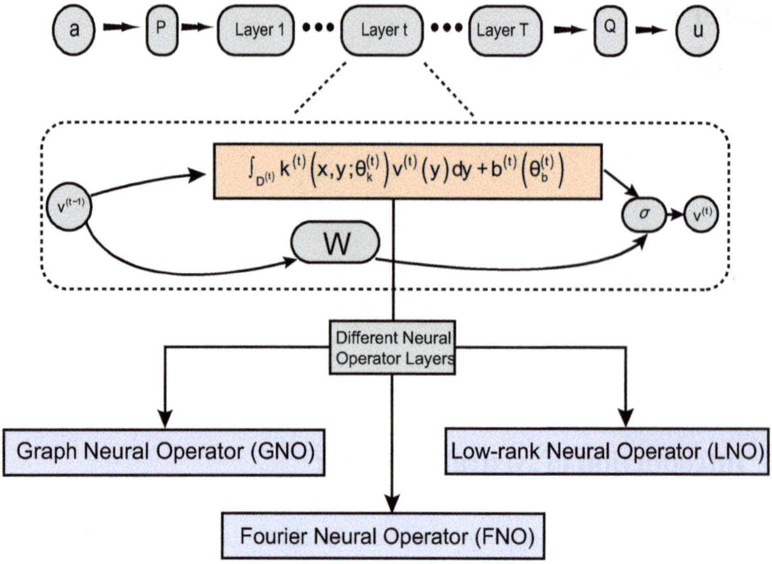

Fig. 6.1 Schematic of Neural Operator. Neural Operator Layer Structure: Graph Neural Operator (GNO) [5], Local Neural Operator (LNO), and Fourier Neural Operator (FNO) [3] can serve as the core of the neural operator architecture [1]

Operator learning is applicable not only to deterministic systems but also to probabilistic models. For example, diffusion-based generative models [4] can learn probability densities over function spaces, making them suitable for tasks such as solving stochastic PDEs.

The design philosophy of operator learning algorithms is illustrated in Fig. 6.1, where the core lies in the kernel integral formulation embedded within neural operator layers:

$$\boldsymbol{H}^{(t)}(\boldsymbol{v}^{(t)}; \boldsymbol{\theta}_k^{(t)})(\boldsymbol{x}) = \int_{D^{(t)}} \boldsymbol{k}^{(t)}(\boldsymbol{x}, \boldsymbol{y}; \boldsymbol{\theta}_k^{(t)})\boldsymbol{v}^{(t)}(\boldsymbol{y})d\boldsymbol{y}, \tag{6.1}$$

where $\boldsymbol{v}^{(t)}(\boldsymbol{y})$ is the input to the t-th neural operator layer, and $\boldsymbol{k}^{(t)}(\boldsymbol{x}, \boldsymbol{y})$ is the kernel function. This kernel plays a role similar to the Green's function in linear PDEs and is approximated by neural networks parameterized by $\boldsymbol{\theta}_k^{(t)}$.

This formulation is referred to as a *kernel integral method* because the kernel operator $\boldsymbol{H}^{(t)}$ acts on a function $\boldsymbol{v}^{(t)}$, transforming it into a new function $\boldsymbol{H}^{(t)}(\boldsymbol{v}^{(t)})(\boldsymbol{x})$ via integration. In the language of mathematical physics, $\boldsymbol{v}^{(t)}(\boldsymbol{y})$ is the original function, $\boldsymbol{H}^{(t)}(\boldsymbol{v}^{(t)})(\boldsymbol{x})$ is the image function, and $\boldsymbol{k}^{(t)}(\boldsymbol{x}, \boldsymbol{y})$ serves as the kernel function facilitating this transformation. The superscript (t) denotes the t-th layer of the neural operator.

The complete forward computation of a neural operator can be expressed as:

$$
\begin{aligned}
G_\theta(\boldsymbol{a}(\boldsymbol{x});\boldsymbol{\theta}) = & \boldsymbol{Q}(\boldsymbol{v}^{(T+1)};\boldsymbol{\theta}_Q)\circ\sigma[\boldsymbol{H}^{(T)}(\boldsymbol{v}^{(T)};\boldsymbol{\theta}_k^{(T)}) + \boldsymbol{W}^{(T)}(\boldsymbol{v}^{(T)};\boldsymbol{\theta}_w^{(T)}) \\
& + \boldsymbol{b}^{(T)}(\boldsymbol{\theta}_b^{(T)})]\circ\cdots\circ\sigma[\boldsymbol{H}^{(1)}(\boldsymbol{v}^{(1)};\boldsymbol{\theta}_k^{(1)}) + \\
& \boldsymbol{W}^{(1)}(\boldsymbol{v}^{(1)};\boldsymbol{\theta}_w^{(1)}) + \boldsymbol{b}^{(1)}(\boldsymbol{\theta}_b^{(1)})]\circ\boldsymbol{P}(\boldsymbol{a}(\boldsymbol{x});\boldsymbol{\theta}_P)
\end{aligned}, \tag{6.2}
$$

where $\boldsymbol{P}$ and $\boldsymbol{Q}$ are the input lifting and output projection operators, respectively, with learnable parameters $\boldsymbol{\theta}_P$ and $\boldsymbol{\theta}_Q$. The components $\boldsymbol{H}^{(t)}$, $\boldsymbol{W}^{(t)}$, and $\boldsymbol{b}^{(t)}$ represent the kernel integral operator, linear transformation, and bias, respectively, each with associated learnable parameters $\boldsymbol{\theta}_k^{(t)}$, $\boldsymbol{\theta}_w^{(t)}$, and $\boldsymbol{\theta}_b^{(t)}$. These components are typically implemented as neural networks.

Different choices of kernel functions lead to various operator learning algorithms, making the design of the kernel function central to the operator learning framework. When the kernel is specifically constructed using the Fourier transform, the resulting algorithm is known as the Fourier Neural Operator (FNO).

There is theoretical evidence that the formulation in Eq. (6.2) satisfies the general approximation properties [6, 7]. As a result, operator learning algorithms based on this formulation possess strong convergence guarantees.

DeepONet [8] also satisfies the general approximation property, and its design is fundamentally rooted in the theoretical framework proposed by [2]. Although DeepONet and FNO are derived from different design philosophies, both exhibit excellent convergence performance in practice.

Among the various operator learning algorithms, DeepONet and FNO are the most representative and widely adopted. Therefore, the following sections will focus on these two methods in detail.

6.1.1 DeepONet

DeepONet is a neural operator learning algorithm based on neural networks. Its mathematical structure is analogous to a Taylor series:

$$
f(\boldsymbol{u})(\boldsymbol{x}) = \sum_{i=1}^{n}\alpha_i(\boldsymbol{u})\phi_i(\boldsymbol{x}). \tag{6.3}
$$

As illustrated in Fig. 6.2, the Trunk net employs neural networks to approximate the basis functions $\phi_i(\boldsymbol{x})$. Trunk net shares a common idea with PINNs, where neural networks fit from the coordinate space to the target function. However, in DeepONet, it is the basis functions $\phi_i(\boldsymbol{x})$ that are fitted by the Trunk net, rather than the target functions. The coefficients before the basis functions $\alpha_i(\boldsymbol{u})$ are fitted by

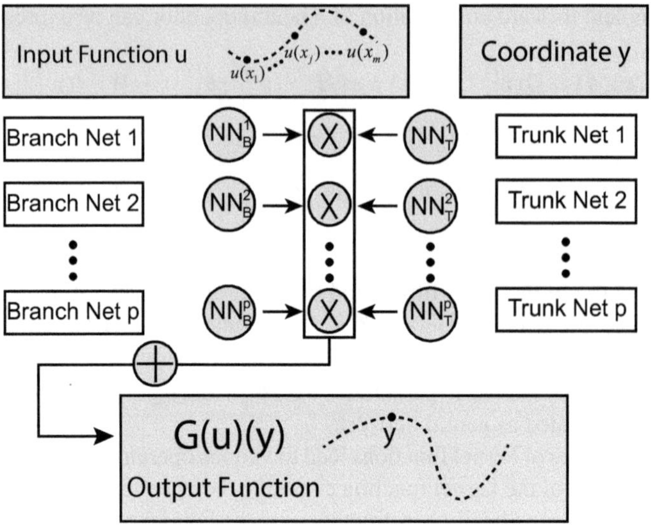

Fig. 6.2 Schematic of DeepONet [8]

the Branch net. It is noteworthy that inputting the same function into the Branch net produces fixed weights $\alpha_i(\boldsymbol{u})$, aligning with the concept of function approximation in numerical analysis. Unlike traditional function approximation algorithms that pre-select basis functions, DeepONet uses neural networks to adaptively select suitable basis functions and weights based on data.

Since the Trunk network is similar to PINNs, we can apply automatic differentiation algorithms in PINNs to construct partial differential operators and obtain new loss functions (strong or energy form). If the Branch network is fixed, DeepONet can be reduced to PINNs.

DeepONet's network structure is constructed according to the theory of universal operator approximation from [2]. The mathematical form of this approximation theory in [2] is:

$$\mid G(\boldsymbol{u})(\boldsymbol{y}) - \sum_{k=1}^{p}\sum_{i=1}^{n} c_i^k \sigma\left(\sum_{j=1}^{m} w_{ij}^k \boldsymbol{u}(\boldsymbol{x}_j) + b_i^k\right)\sigma(\boldsymbol{W}^k \cdot \boldsymbol{y} + \boldsymbol{B}^k) \mid < \epsilon. \tag{6.4}$$

DeepONet replaces the terms in the above equation with neural networks:

$$\begin{aligned}
NN_B^k(\boldsymbol{u}(\boldsymbol{x}); \boldsymbol{\theta}_{NN}^B) &= \sum_{i=1}^{n} c_i^k \sigma\left(\sum_{j=1}^{m} w_{ij}^k \boldsymbol{u}(\boldsymbol{x}_j) + b_i^k\right) \\
NN_T^k(\boldsymbol{y}; \boldsymbol{\theta}_{NN}^T) &= \sigma(\boldsymbol{W}^k \cdot \boldsymbol{y} + \boldsymbol{B}^k) \\
NN_O(\boldsymbol{y}; \boldsymbol{u}) &= \sum_{k=1}^{p} NN_B^k(\boldsymbol{u}(\boldsymbol{x}); \boldsymbol{\theta}_{NN}^B) \cdot NN_T^k(\boldsymbol{y}; \boldsymbol{\theta}_{NN}^T)
\end{aligned}, \tag{6.5}$$

where NN_B^k, NN_T^k, and NN_O are respectively the Branch network, Trunk network, and the output of DeepONet. $\boldsymbol{u}(\boldsymbol{x})$ can be approximated from sensors $\boldsymbol{x}$ (discrete points approximating a continuous function). For instance, if the input $\boldsymbol{u}(\boldsymbol{x})$ of NN_B is the gravitational potential energy of a three-dimensional cubic material over $[0, 1]^3$, then the gravitational potential energy field within $[0, 1]^3$ might be sampled at 0.01 intervals taking 101^3 points. Data structure of $\boldsymbol{u}(\boldsymbol{x})$ can vary, such as data structures similar to images. Therefore, we can use network structures for the Branch network that are appropriate for the type of data, such as CNNs for local features similar to image data, and RNNs for time-related features. It is important that the network structure of the Branch network must consider the particularities of the problem to choose the most appropriate network structure. Thus, $\boldsymbol{u}(\boldsymbol{x})$ determines the output of the Branch network in DeepONet, which is related to the weights $\alpha_i(\boldsymbol{u})$ in Eq. (6.3) and is independent of the coordinates of interest. Similarly, the coordinates fully determine the output of the Trunk network, precisely defining the basis functions $\phi_i(\boldsymbol{x})$ in Eq. (6.3). Therefore, DeepONet shares many similarities with traditional function approximation. It can be said that DeepONet can adaptively learn basis functions and weights based on data. For more details and extensions on DeepONet, see [8].

Listings 12 and 13 show examples of the DeepONet architecture, corresponding to the Branch network and Trunk network in DeepONet, respectively.

6.1.2 FNO: Fourier Neural Operator

The Fourier Neural Operator (FNO) is an operator learning algorithm introduced by Li et al. [3], which has received considerable attention. The initially proposed FNO is a completely data-driven model. Even earlier than FNO, the same first author, Li et al. [5], used graph neural networks to learn relationships between field variables (GNO), but FNO demonstrated better performance than GNO. The essence of FNO involves integrating Fourier transforms into operator learning, representing Eq. (6.1) with a Fourier transform:

$$\boldsymbol{H}^{(t)}(\boldsymbol{v}^{(t)}; \boldsymbol{\theta}_k^{(t)})(\boldsymbol{x}) = \int_{D^{(t)}} \boldsymbol{k}^{(t)}(\boldsymbol{x}, \boldsymbol{y}; \boldsymbol{\theta}_k^{(t)})\boldsymbol{v}^{(t)}(\boldsymbol{y})d\boldsymbol{y} = \mathcal{F}^{-1} \circ \mathrm{Re} \circ \mathcal{F}(\boldsymbol{v}^{(t)}(\boldsymbol{x})), \tag{6.6}$$

where $\mathcal{F}$ and $\mathcal{F}^{-1}$ are the Fourier transform and its inverse, respectively, $\Re$ is a linear transformation. To clarify the process of the FNO algorithm, we combine the framework and examples of the FNO algorithm. The framework of the FNO

```
class Branch_net(torch.nn.Module):
    def __init__(self, D_in, H, D_out):
        """
        DeepONet: Branch net
        """
        super(Branch_net, self).__init__()
        self.linear1 = torch.nn.Linear(D_in, H)
        self.linear2 = torch.nn.Linear(H, H)
        self.linear3 = torch.nn.Linear(H, H)
        self.linear4 = torch.nn.Linear(H, D_out)

        torch.nn.init.normal_(self.linear1.weight,\
            mean=0, std=np.sqrt(2/(D_in+H)))
        torch.nn.init.normal_(self.linear2.weight,\
            mean=0, std=np.sqrt(2/(H+H)))
        torch.nn.init.normal_(self.linear3.weight,\
            mean=0, std=np.sqrt(2/(H+H)))
        torch.nn.init.normal_(self.linear4.weight,\
            mean=0, std=np.sqrt(2/(H+D_out)))
        torch.nn.init.normal_(self.linear1.bias, mean=0, std=1)
        torch.nn.init.normal_(self.linear2.bias, mean=0, std=1)
        torch.nn.init.normal_(self.linear3.bias, mean=0, std=1)
        torch.nn.init.normal_(self.linear4.bias, mean=0, std=1)
    def forward(self, x):
        y1 = torch.tanh(self.linear1(x))
        y2 = torch.tanh(self.linear2(y1))
        y3 = torch.tanh(self.linear3(y2))
        y = self.linear4(y3)
        return y
```

Listing 12: Example of the Branch network in DeepONet

algorithm, as shown in Fig. 6.3, considers the one-dimensional Burgers equation, commonly used as a benchmark due to its space-time and nonlinear characteristics:

$$\begin{cases} \text{Domain control equation}: \frac{\partial u}{\partial t} + u\frac{\partial u}{\partial x} = \nu\frac{\partial^2 u}{\partial x^2} \quad x \in [-1, 1],\ t \geq 0 \\ \text{Boundary conditions}: u(-1, t) = u(1, t) = 0 \\ \text{Initial condition: } u(x, 0) = u_0(x) \end{cases}, \tag{6.7}$$

where u is the field variable to be solved, ν is the viscosity coefficient, and the boundary conditions are fixed. Once the initial condition u_0 is determined, the solution u can be uniquely determined. Thus, in this problem, the input is the initial condition $u_0 \in \mathbb{R}^{d_i}$ (i.e., $a(x)$ in [3]), and the output is $u(x, t_1)$ at a future time t_1. The FNO algorithm process first uses a fully connected neural network $\boldsymbol{P}$ to elevate the dimension of the input $a(x)$ to d_c (i.e., $P : \mathbb{R} \to \mathbb{R}^{d_c}$, d_c similar to channels in computer vision):

$$\boldsymbol{v}^{(0)}(x) = \boldsymbol{P}(a(x)) \in \mathbb{R}^{d_i * d_c}. \tag{6.8}$$

```python
class Trunk_net(torch.nn.Module):
    def __init__(self, D_in, H, D_out):
        """
        DeepONet: Trunk Net
        """
        super(Trunk_net, self).__init__()
        self.linear1 = torch.nn.Linear(D_in, H)
        self.linear2 = torch.nn.Linear(H, H)
        self.linear3 = torch.nn.Linear(H, H)
        self.linear4 = torch.nn.Linear(H, D_out)

        torch.nn.init.normal_(self.linear1.weight,\
            mean=0, std=np.sqrt(2/(D_in+H)))
        torch.nn.init.normal_(self.linear2.weight,\
            mean=0, std=np.sqrt(2/(H+H)))
        torch.nn.init.normal_(self.linear3.weight,\
            mean=0, std=np.sqrt(2/(H+H)))
        torch.nn.init.normal_(self.linear4.weight,\
            mean=0, std=np.sqrt(2/(H+D_out)))
        torch.nn.init.normal_(self.linear1.bias, mean=0, std=1)
        torch.nn.init.normal_(self.linear2.bias, mean=0, std=1)
        torch.nn.init.normal_(self.linear3.bias, mean=0, std=1)
        torch.nn.init.normal_(self.linear4.bias, mean=0, std=1)

    def forward(self, x):
        """
        Input: the coordinate
        Output: the basis of the DeepONet
        """
        y1 = torch.tanh(self.linear1(x))
        y2 = torch.tanh(self.linear2(y1))
        y3 = torch.tanh(self.linear3(y2))
        y = self.linear4(y3)
        return y
```

Listing 13: Example of the Trunk network in DeepONet

Next, $v^{(0)}(x)$ is sent into Fourier layer, where each channel of $v^{(0)}(x)$ undergoes a Fourier transform, transforming into:

$$\boldsymbol{f}^{(1)}(x) = \mathcal{F}(\boldsymbol{v}^{(0)}(x)) \in \mathbb{C}^{d_i * d_c}. \tag{6.9}$$

High frequencies are filtered out, retaining low frequencies, and cutting off the frequency at $d_m (1 \leq d_m \leq d_i)$, thus the data structure transforms into $f^{(1)}(x) \in \mathbb{C}^{d_m \times d_c}$. A linear transformation $\Re$ is applied to each frequency of all channels:

$$\boldsymbol{f}_L^{(1)}(x) = \mathrm{Re}(\boldsymbol{f}^{(1)}(x)) \in \mathbb{C}^{d_m * d_c}. \tag{6.10}$$

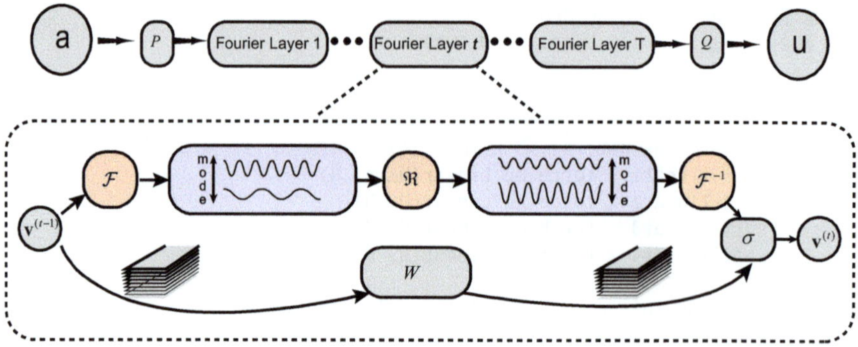

Fig. 6.3 Schematic of Fourier Neural Operator (FNO) [3]

Since different frequencies use different linear transformations, there are d_m linear transformation matrices, and the linear transformation is from d_c to d_c, so the linear transformation matrix is $\mathbb{R}^{d_c \times d_c \times d_m}$. After the linear transformation $\mathfrak{R}$, the data's dimensional structure is not changed. Since the next step involves Fourier inverse transformation, high frequencies removed by cutting are replaced by zeros. Fourier inverse transformation is performed on $\boldsymbol{f}_L^{(1)}(x)$:

$$\boldsymbol{v}_f^{(1)} = \mathcal{F}^{-1}(\boldsymbol{f}_L^{(1)}(x)) \in \mathbb{R}^{d_i * d_c}. \tag{6.11}$$

The advantage of the linear transformation $\mathfrak{R}$ is that it learns the mapping relationships between data in the Fourier frequency space while filtering out high frequencies. $\mathfrak{R}$ is a special form of regularization that reduces overfitting and enhances the model's generalization capability.

Furthermore, using the idea of residual networks from [9], another linear transformation is directly applied to the channels of $v^{(0)}(x)$, the same linear transformation for all different frequencies, only one linear transformation, i.e., $\boldsymbol{W} \in \mathbb{R}^{d_c \times d_c}$:

$$\boldsymbol{v}_L^{(1)} = \boldsymbol{W}(\boldsymbol{v}^{(0)}(x)) \in \mathbb{R}^{d_i * d_c}. \tag{6.12}$$

$\boldsymbol{W}$ uses the idea of residual networks to reintroduce high-frequency components filtered out by the Fourier transform back into the network for learning, preventing a decrease in network prediction performance due to the high frequencies filtered out by the Fourier transform. The results of the Fourier transform and the linear transformation are added together to obtain:

$$\boldsymbol{v}^{(1)}(x) = \sigma(\boldsymbol{v}_f^{(1)} + \boldsymbol{v}_L^{(1)} + \boldsymbol{b}^{(1)}) \in \mathbb{R}^{d_i * d_c}, \tag{6.13}$$

where $\boldsymbol{b}^{(1)}$ is the bias. Adding a nonlinear transformation σ not only increases the network's expressive capability but also further enhances the high-frequency fitting capability. The Fourier layer is the computational core of FNO, and the above operations are repeated T times to obtain $v^{(T)}(x)$. Finally, the dimension d_c is reduced to the target dimension by $\boldsymbol{Q} : \mathbb{R}^{d_c} \rightarrow \mathbb{R}$.

The two major advantages of FNO are, first, since the Fourier transform filters out high-frequency modes, it can enhance the model's speed and generalization ability, reducing overfitting; second, discretization-invariance, since all operations in FNO are unrelated to the data's mesh and are based on point-wise operations, so it has the advantage of discretization-invariance. The initially proposed FNO has two major flaws. First, the Fourier Neural Operator (FNO) employs the fast Fourier transform, which requires the input data to be defined over a regular domain. Although it is feasible to enclose an irregular domain within a larger regular one, this approach often yields suboptimal results. To address this, Li et al. [10] introduced Geo-FNO, specifically designed to handle irregular domains effectively. Second, while FNO requires data points to be positioned on a uniform mesh, points on an uneven mesh must be converted into a uniform mesh using interpolation. Thus, the error of interpolation decreases the accuracy of FNO.

Listing 14 presents the implementation details of the 3D Fourier layer, which is a fundamental component of the 3D Fourier Neural Operator (FNO). Listing 15 illustrates the complete pipeline of the 3D FNO model.

For the comparison between FNO and DeepONet, Lu et al. [11] systematically analyzed both methods. They introduced enhancements to FNO to handle mappings of different dimensionalities and complex geometries, while also improving DeepONet to accelerate training and enhance accuracy, supplemented by theoretical comparisons.

6.2 Application of DeepONet in the Deep Energy Method

In this chapter, we specifically demonstrate how DeepONet can be applied to the Deep Energy Method (DEM). The starting point of this idea is to fully reuse the existing computational results, rather than discarding them, in order to improve the computational efficiency of DEM.

In this demo, we reformulate the loss function from the principle of minimum potential energy to the principle of complementary energy. The principles of minimum potential energy and complementary energy are two fundamental variational principles in linear elasticity [12].

Based on this idea, Wang et al. [13] proposed the Deep Complementary Energy Method (DCEM), which essentially replaces the loss function based on the minimum potential energy principle in DEM with one based on the minimum complementary energy principle. Correspondingly, the primary variable changes from the displacement field in DEM to the stress function in DCEM.

In DCEM, a coupling with DeepONet has been implemented, namely, DCEM-O. In the following, we will focus on the methodology and implementation of DCEM-O.

```
class SpectralConv3d(nn.Module):
    def __init__(self, in_channels, out_channels, modes1, modes2, modes3):
        super(SpectralConv3d, self).__init__()
        """
        3D Fourier layer. It does FFT, linear transform, and Inverse FFT.
        """
        self.in_channels = in_channels
        self.out_channels = out_channels
        self.modes1 = modes1
        self.modes2 = modes2
        self.modes3 = modes3

        self.scale = (1 / (in_channels * out_channels))
        self.weights1 = nn.Parameter(self.scale * torch.rand(in_channels,\
            out_channels, self.modes1, self.modes2,\
            self.modes3, dtype=torch.cfloat))
        self.weights2 = nn.Parameter(self.scale * torch.rand(in_channels,\
            out_channels, self.modes1, self.modes2,\
            self.modes3, dtype=torch.cfloat))
        self.weights3 = nn.Parameter(self.scale * torch.rand(in_channels,\
            out_channels, self.modes1, self.modes2,\
            self.modes3, dtype=torch.cfloat))
        self.weights4 = nn.Parameter(self.scale * torch.rand(in_channels,\
            out_channels, self.modes1, self.modes2,\
            self.modes3, dtype=torch.cfloat))
    # Complex multiplication
    def compl_mul3d(self, input, weights):
        return torch.einsum("bixyz,ioxyz->boxyz", input, weights)
    def forward(self, x):
        batchsize = x.shape[0]
        x_ft = torch.fft.rfftn(x, dim=[-3,-2,-1])
        # Multiply relevant Fourier modes
        out_ft = torch.zeros(batchsize, self.out_channels,\
            x.size(-3), x.size(-2), x.size(-1)//2 + 1, \
            dtype=torch.cfloat, device=x.device)
        out_ft[:, :, :self.modes1, :self.modes2, :self.modes3] = \
            self.compl_mul3d(x_ft[:, :, :self.modes1, :self.modes2,\
            :self.modes3], self.weights1)
        out_ft[:, :, -self.modes1:, :self.modes2, :self.modes3] = \
            self.compl_mul3d(x_ft[:, :, -self.modes1:, :self.modes2,\
            :self.modes3], self.weights2)
        out_ft[:, :, :self.modes1, -self.modes2:, :self.modes3] = \
            self.compl_mul3d(x_ft[:, :, :self.modes1, -self.modes2:, \
            :self.modes3], self.weights3)
        out_ft[:, :, -self.modes1:, -self.modes2:, :self.modes3] = \
            self.compl_mul3d(x_ft[:, :, -self.modes1:, -self.modes2:,\
            :self.modes3], self.weights4)
        #Return to physical space
        x = torch.fft.irfftn(out_ft, s=(x.size(-3), x.size(-2), x.size(-1)))
        return x
```

Listing 14: Implementation of the 3D Fourier layer

```python
class FNO3d(nn.Module):
    def __init__(self, modes1, modes2, modes3, width):
        super(FNO3d, self).__init__()
        self.modes1 = modes1
        self.modes2 = modes2
        self.modes3 = modes3
        self.width = width
        self.fc0 = nn.Linear(Input_dim, self.width)

        self.conv0 = SpectralConv3d(self.width, self.width,\
            self.modes1, self.modes2, self.modes3)
        self.conv1 = SpectralConv3d(self.width, self.width,\
            self.modes1, self.modes2, self.modes3)
        self.conv2 = SpectralConv3d(self.width, self.width,\
            self.modes1, self.modes2, self.modes3)
        self.conv3 = SpectralConv3d(self.width, self.width,\
            self.modes1, self.modes2, self.modes3)
        self.w0 = nn.Conv3d(self.width, self.width, 1)
        self.w1 = nn.Conv3d(self.width, self.width, 1)
        self.w2 = nn.Conv3d(self.width, self.width, 1)
        self.w3 = nn.Conv3d(self.width, self.width, 1)
        self.bn0 = torch.nn.BatchNorm3d(self.width)
        self.bn1 = torch.nn.BatchNorm3d(self.width)
        self.bn2 = torch.nn.BatchNorm3d(self.width)
        self.bn3 = torch.nn.BatchNorm3d(self.width)

        self.fc1 = nn.Linear(self.width, 64)
        self.fc2 = nn.Linear(64, 18)
    def forward(self, x):
        x = self.fc0(x)

        x1 = self.conv0(x)
        x2 = self.w0(x)
        x = x1 + x2
        x = F.gelu(x)

        x1 = self.conv1(x)
        x2 = self.w1(x)
        x = x1 + x2
        x = F.gelu(x)

        x1 = self.conv2(x)
        x2 = self.w2(x)
        x = x1 + x2
        x = F.gelu(x)

        x1 = self.conv3(x)
        x2 = self.w3(x)
        x = x1 + x2

        x = self.fc1(x)
        x = F.gelu(x)
        x = self.fc2(x)
        return x
```

Listing 15: Implementation of the complete 3D FNO model

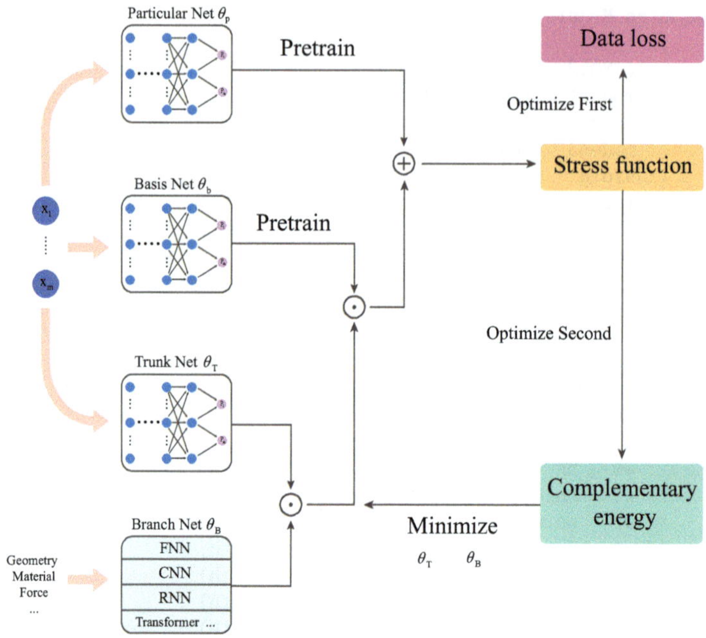

Fig. 6.4 Schematic of the deep operator energy method based on the principle of minimum complementary energy (DCEM-O)

The idea of DCEM-O is to use the DeepONet framework to help DCEM make operator learning (DeepONet serves as the framework for our operator learning), as mentioned in Sect. 6.1.1. Trunk net can build partial differential operators to construct the physical loss.

Figure 6.4 illustrates the framework of DCEM-O. The training process follows a sequential procedure: the operator is first learned from existing data, i.e., the physics-based complementary energy functional is not constructed initially. After the data-driven training stage, the model is further optimized based on the principle of complementary energy.

In the DCEM-O framework, the input to the branch network may include different geometries, material properties, and force boundary conditions (other field variables can also be included), while the output corresponds to existing high-fidelity solutions of the stress function, such as those obtained from stress-function-based finite element methods. Finally, the parameters of both the trunk and branch networks are optimized according to the principle of minimum complementary energy.

The key to operator learning with physical laws lies in effectively integrating physics-based principles into data-driven frameworks. Numerous approaches have been proposed to bridge the gap between physical models and data-driven methods [14–16].

In our study, the physical loss is constructed using automatic differentiation (AD), rather than numerical differentiation. This choice is motivated by the fact that finite difference methods require a fine, uniform grid; without such a grid, numerical errors in derivatives may amplify and adversely affect the solution accuracy. Therefore, AD is preferred for constructing physical loss terms.

Hybrid optimization strategies [14–16] combine data loss and physical loss into a unified objective function with tunable hyperparameters. A typical formulation is:

$$\begin{cases} \mathcal{L}(\boldsymbol{\theta}) = \mathcal{L}_{\text{data}}(\boldsymbol{\theta}) + \lambda\, \mathcal{L}_{\text{phy}}(\boldsymbol{\theta}) \\ \mathcal{L}_{\text{data}}(\boldsymbol{\theta}) = \frac{1}{IJ} \sum_{j=1}^{J} \sum_{i=1}^{I} \left| NN_O(\boldsymbol{y}^{(i)}, \boldsymbol{u}^{(j)}; \boldsymbol{\theta}) - \boldsymbol{u}^{(j)}(\boldsymbol{y}^{(i)}) \right|^2 \\ \mathcal{L}_{\text{phy}}(\boldsymbol{\theta}) = \frac{1}{JK} \sum_{j=1}^{J} \sum_{k=1}^{K} \left| \text{Phy}(NN_O(\boldsymbol{x}^{(k)}, \boldsymbol{u}^{(j)}; \boldsymbol{\theta})) \right|^2 \end{cases}, \tag{6.14}$$

where the details of Eq. (6.14) are provided in Sect. 6.1.1. This approach blends the data-driven (operator) loss with the physics-informed loss using a hyperparameter λ.

However, in this chapter, we adopt a sequential optimization strategy instead of the hybrid one. This decision is based on the difficulty in selecting an appropriate value for the hyperparameter λ. Specifically, we first minimize the data loss and then perform a second-stage optimization using the physics-based complementary energy principle.

Next, we consider a specific example that combines DeepONet with DCEM: the circular tube.

The circular tube problem is relevant to the study of thick-walled pressure vessels, which are widely used as containers for gas and other substances in engineering applications [17].

In general, if all boundary conditions are traction (force) boundary conditions, the admissibility of the displacement field is not required in DEM. In this example, however, all boundary conditions are displacement boundary conditions. As a result, the admissibility of the stress function is not required in DCEM.

Although DEM can also be applied to problems with fully prescribed displacement boundary conditions, it requires strong assumptions about the admissible displacement fields to ensure compliance with boundary conditions and variational principles.

As shown in Fig. 6.5, we utilize the analytical solution given in Eq. (6.15) to convert the fully pressure-prescribed boundary conditions into fully displacement-prescribed boundary conditions. This transformation enables a direct comparison between DEM and DCEM under identical displacement boundary conditions, thereby highlighting the respective advantages and limitations of different energy principles.

In this setting, DCEM does not require the assumption of an admissible stress function, since no force boundary conditions are present. We employ the displacement solution derived from the classical Lame's formula to construct the

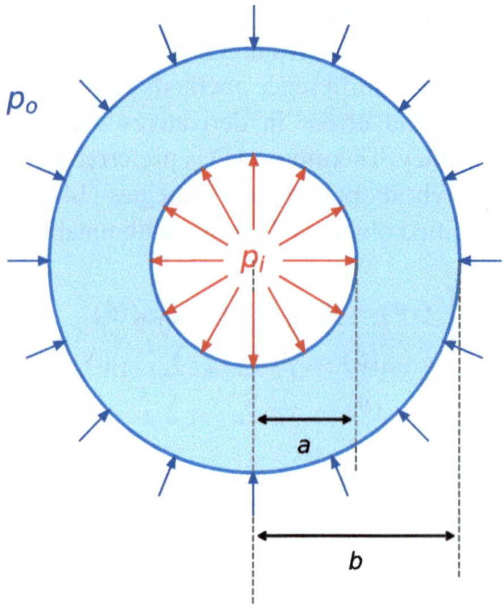

Fig. 6.5 Schematic diagram of the circular tube: parameter values are $p_o = 10$, $p_i = 5$, $a = 0.5$, $b = 1.0$, Young's modulus $E = 1000$, and Poisson's ratio $\nu = 0.3$

displacement boundary conditions. The analytical expressions for the Lame stress and displacement fields are given by:

$$\begin{aligned} \sigma_r &= \frac{a^2}{b^2 - a^2}\left(1 - \frac{b^2}{r^2}\right) p_i - \frac{b^2}{b^2 - a^2}\left(1 - \frac{a^2}{r^2}\right) p_o \\ \sigma_\theta &= \frac{a^2}{b^2 - a^2}\left(1 + \frac{b^2}{r^2}\right) p_i - \frac{b^2}{b^2 - a^2}\left(1 + \frac{a^2}{r^2}\right) p_o \\ u_r &= \frac{1}{E}\left[\frac{(1-\nu)(a^2 p_i - b^2 p_o)}{b^2 - a^2} r + \frac{(1+\nu)a^2 b^2 (p_i - p_o)}{b^2 - a^2}\frac{1}{r}\right] \\ u_\theta &= 0, \end{aligned} \tag{6.15}$$

where a and b denote the inner and outer radii, respectively; E is the Young's modulus; ν is the Poisson's ratio; and p_i and p_o are the uniform internal and external pressures. Note that compressive stress is considered positive in this formulation.

To apply DCEM to this problem, the complementary strain energy and complementary potential must be expressed in terms of the stress function. For an axisymmetric problem in polar coordinates, the stress components in terms of the stress function ϕ are:

$$\begin{aligned} \sigma_r &= \frac{1}{r}\frac{\partial \phi}{\partial r} \\ \sigma_\theta &= \frac{\partial^2 \phi}{\partial r^2}. \end{aligned} \tag{6.16}$$

Using the constitutive relationships in polar coordinates, the strain components can be expressed as:

$$\begin{aligned}\varepsilon_r &= \frac{1}{E}(\sigma_r - \nu\sigma_\theta) = \frac{1}{E}\left(\frac{1}{r}\frac{\partial\phi}{\partial r} - \nu\frac{\partial^2\phi}{\partial r^2}\right)\\ \varepsilon_\theta &= \frac{1}{E}(\sigma_\theta - \nu\sigma_r) = \frac{1}{E}\left(\frac{\partial^2\phi}{\partial r^2} - \nu\frac{1}{r}\frac{\partial\phi}{\partial r}\right)\end{aligned} \tag{6.17}$$

The complementary strain energy W_c in terms of ϕ is then given by:

$$\begin{aligned}W_c &= \int_\Omega \frac{1}{2}\sigma_{ij}\varepsilon_{ij}\, d\Omega = \int_\Omega \frac{1}{2}(\sigma_r\varepsilon_r + \sigma_\theta\varepsilon_\theta)\, 2\pi r\, dr\\ &= \int_\Omega \frac{1}{2}\left[\frac{1}{r}\frac{\partial\phi}{\partial r}\frac{1}{E}\left(\frac{1}{r}\frac{\partial\phi}{\partial r} - \nu\frac{\partial^2\phi}{\partial r^2}\right) + \frac{\partial^2\phi}{\partial r^2}\frac{1}{E}\left(\frac{\partial^2\phi}{\partial r^2} - \nu\frac{1}{r}\frac{\partial\phi}{\partial r}\right)\right] 2\pi r\, dr\end{aligned} \tag{6.18}$$

The complementary potential V_c is expressed as:

$$\begin{aligned}V_c &= u_r\sigma_r 2\pi r|_{r=b} - u_r\sigma_r 2\pi r|_{r=a}\\ &= u_r\frac{\partial\phi}{\partial r}2\pi\bigg|_{r=b} - u_r\frac{\partial\phi}{\partial r}2\pi\bigg|_{r=a}\end{aligned} \tag{6.19}$$

In this full-displacement boundary condition example, we compare DCEM-O (operator learning based on DCEM) with the original DCEM. The input to the branch network consists of p_i and p_o (pressure boundary conditions). Since no force boundary conditions are present, no basis or particular solutions are required.

The output of DCEM-O is the predicted Airy stress function ϕ, which is represented as the product of the branch and trunk networks:

$$\varphi = NN_B(p_i, p_o; \boldsymbol{\theta}_B) \cdot NN_T(r; \boldsymbol{\theta}_T). \tag{6.20}$$

The trunk network consists of 6 hidden layers with 30 neurons per layer, while the branch network contains 3 hidden layers with 30 neurons per layer. A total of 10,000 training samples are generated, with inputs covering variations in p_i, p_o (boundary conditions), ν (material property), and b (geometry). Each parameter ($p_i \in [4.0, 6.0]$, $p_o \in [9.0, 11.0]$, $\nu \in [0.3, 0.49]$, $b \in [0.9, 1.1]$) is uniformly discretized into 10 points, resulting in 10^4 input combinations.

The generation of the DeepONet training dataset is shown in Listing 16.

Listing 17 presents the training code for DeepONet in the DCEM-O framework. Each data point takes approximately 2.46 seconds to train. The output of DCEM-O is the Airy stress function ϕ, and the ground truth training data are obtained either from the analytical solution or from high-fidelity numerical simulations, such as the finite element method based on the stress function [18].

```
# ==========================================================================
# DeepONet dataset: we try a/b ratio is 1.0 for the test, and the other datasets
# for training set
# ==========================================================================
Po_list_train = list(np.linspace(9, 11, 10)) # for DeepONet training
Pi_list_train = list(np.linspace(4, 6, 10)) # for DeepONet training
poisson_list_train = list(np.linspace(0.3, 0.49, 10)) # for DeepONet training
b_list_train = list(np.linspace(0.9, 1.1, 10)) # for DeepONet training

# ==========================================================================
# DeepONet dataset: we try a/b ratio is 1.0 for the test, and the other datasets
# for training set
# ==========================================================================
stress_data_set = []
for Pi_e in Pi_list_train:
    for Po_e in Po_list_train:
        for poisson_e in poisson_list_train:
            for b_e in b_list_train:
                Ura_e = 1/E*((1-poisson_e)*(a**2*Pi_e-b_e**2*Po_e)\
                    /(b_e**2-a**2)*a + (1+poisson_e)*(a**2*b_e**2)*(Pi_e-Po_e)\
                    /(b_e**2-a**2)/a)
                Urb_e = 1/E*((1-poisson_e)*(a**2*Pi_e-b_e**2*Po_e)\
                    /(b_e**2-a**2)*b_e + (1+poisson_e)*(a**2*b_e**2)*(Pi_e-Po_e)\
                    /(b_e**2-a**2)/b_e)
                stress_data_set.append([Pi_e, Po_e, Ura_e, Urb_e, poisson_e, b_e])
stress_data_set = np.array(stress_data_set)
point_data_set = np.zeros([len(stress_data_set), N_DeepONet_t])
fai_data_set = np.zeros([len(stress_data_set), N_DeepONet_t])
for i, Pi_Po in enumerate(stress_data_set):
    Pi_e = Pi_Po[0]
    Po_e = Pi_Po[1]
    b_e = Pi_Po[5]

    r = np.linspace(a, b, N_DeepONet_t)
    point_data_set[i] = r

    fai_exact = a**2/(b_e**2-a**2)*(r**2/2-b_e**2*np.log(r))*\
                Pi_e-b_e**2/(b_e**2-a**2)\
        *(r**2/2-a**2*np.log(r))*Po_e
    fai_data_set[i] = fai_exact
stress_data_set = stress_data_set[:, 2:] # delete the stress and leave the dis
```

Listing 16: The training dataset of DCEM-O

The analytical solution for the Airy stress function in the circular tube problem is given by:

$$\phi = \frac{a^2}{b^2 - a^2}\left[\frac{r^2}{2} - b^2 \ln(r)\right] p_i - \frac{b^2}{b^2 - a^2}\left[\frac{r^2}{2} - a^2 \ln(r)\right] p_o. \tag{6.21}$$

Note that the test dataset used in our study ($p_i = 5$, $p_o = 10$, Poisson's ratio = 0.5, and $b = 1.0$) lies outside the training range. This test is of particular interest because a Poisson's ratio of 0.5 corresponds to an incompressible material, which allows us to examine the extrapolation capability of DCEM-O.

```python
def pred(xy, stress):
    '''



    Parameters
    ----------
    xy : tensor
        the coordinate of the input tensor.
    stress : Pi and Po

    Returns
    the prediction of fai

    '''

    pred_fai = torch.sum((model_Branch_net(stress)*model_Trunk_net(xy)),\
            1,keepdims = True)

    return pred_fai

# =============================================================================
#    DeepONet training
# =============================================================================
model_Branch_net = Branch_net(4, 30, 20).cuda()
model_Trunk_net = Trunk_net(1, 30, 20).cuda()
criterion = torch.nn.MSELoss()
optim = torch.optim.Adam(params=chain(model_Branch_net.parameters(), \
        model_Trunk_net.parameters()), lr= 0.001)
loss_array_DeepONet = {}
for iter_all in range(iter_data):
    for i, Pi_Po in enumerate(stress_data_set):
        loss_array_DeepONet[i] = []
        stress_eve = torch.tensor(stress_data_set[i]).cuda()
        xy_set = torch.tensor(point_data_set[i][:, np.newaxis]).cuda()
        fai_set = torch.tensor(fai_data_set[i][:, np.newaxis]).cuda()
        start = time.time()
        for epoch in range(nepoch_DeepONet):
            def closure():
                #
                pred_fai = pred(xy_set, stress_eve)
                loss = criterion(pred_fai, fai_set)
                optim.zero_grad()
                loss.backward(retain_graph=True)
                loss_array_DeepONet[i].append(loss.data.cpu())

                if epoch%10==0:
                    print('whole iter: %i, pi : %f, po : %f,    epoch : \
                    %i, the loss : %f' % \
                          (iter_all, Pi_Po[0], Pi_Po[1], epoch, loss.data))
                return loss
            optim.step(closure)
```

Listing 17: The training process of DCEM-O

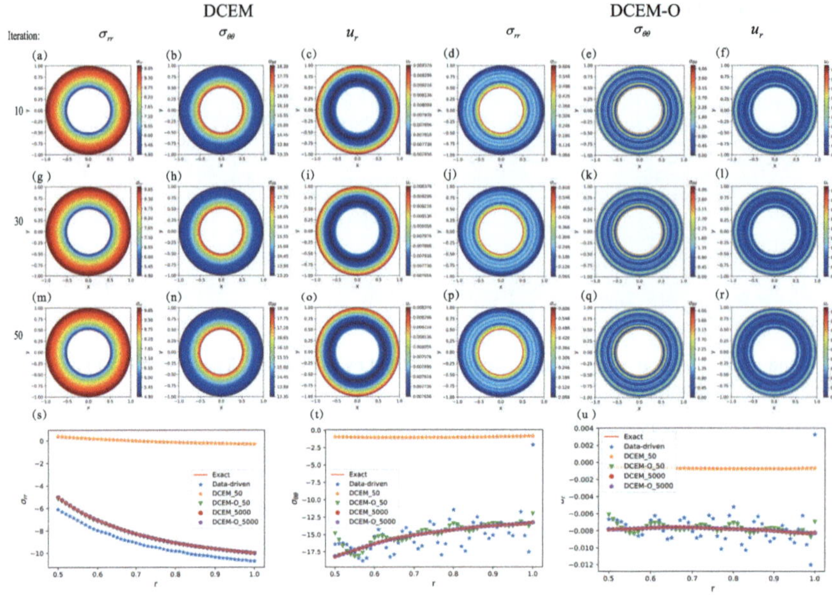

Fig. 6.6 Comparison of DCEM and DCEM-O in the Airy stress function of the circular tube: (a–r) show the absolute error after 10, 30, and 50 iterations; (a, g, m, d, j, p) represent σ_{rr}; (b, h, n, e, k, q) represent $\sigma_{\theta\theta}$; (c, i, o, f, l, r) represent u_r; results from DCEM are shown in (a, b, c, g, h, i, m, n, o), and results from DCEM-O are shown in (d, e, f, j, k, l, p, q, r); subfigures (s), (t), and (u) compare the predictions of σ_{rr}, $\sigma_{\theta\theta}$, and u_r after 50 and 5000 iterations, respectively. The label "data-driven" indicates DCEM-O predictions before fine-tuning with physical loss

Figure 6.6 presents a comparison between DCEM and DCEM-O in terms of solution accuracy. The absolute error of DCEM-O is consistently lower than that of DCEM at the same iteration number. Subfigures (s), (t), and (u) show that DCEM-O converges faster than DCEM in the early training stages, and both methods converge to the exact solution after 5000 iterations. Due to differences in neural network architectures, the average time per training epoch is 0.011 seconds for DCEM-O and 0.0058 seconds for DCEM.

Listing 18 provides the implementation for testing DCEM-O.

Although the previous example demonstrates the effectiveness of DCEM-O on a single test case, we further perform statistical experiments to assess its generalization ability. Specifically, we test DCEM-O across 18 equally spaced values of Poisson's ratio in the range $\nu \in [0.31, 0.32, \ldots, 0.48]$. Note that these values lie outside the set of training Poisson's ratios, which are sampled at 10 equally spaced points in the range [0.3, 0.49], including the endpoints.

Both DCEM and DCEM-O are evaluated under the same stopping criterion: iterations terminate when the relative $\mathcal{L}_2$ error reaches 1%. The results in Fig. 6.7 indicate that DCEM-O achieves convergence significantly faster than DCEM for most values of Poisson's ratio.

```python
optim = torch.optim.Adam(params=chain(model_Branch_net.parameters(),\
                            model_Trunk_net.parameters()), lr= 0.001)
step_scheduler = torch.optim.lr_scheduler.StepLR(optim, step_size=\
                                2000, gamma=0.1)
stress = torch.tensor([Ura, Urb, nu=0.5, b=1.0]).cuda()
Xf = dom_data(dom_num)

for epoch in range(nepoch):
    def closure():
        #
        r = Xf
        fai = pred(r, stress)
        dfaidr = grad(fai, r, torch.ones(r.size()[0], 1).cuda(), \
                        retain_graph=True, create_graph=True)[0]
        dfaidrr = grad(dfaidr, r, torch.ones(r.size()[0], 1).cuda(),\
                        retain_graph=True, create_graph=True)[0]

        J_dom = simpson_int(0.5/E*(1/r**2*dfaidr**2 + dfaidrr**2 \
                            - 2*nu/r*dfaidr*dfaidrr)*2*np.pi*r, r)
        # calculate the complementary external energy
        ra = torch.tensor([a], requires_grad=True, device='cuda')\
                            .unsqueeze(1)
        rb = torch.tensor([b], requires_grad=True, device='cuda')\
                            .unsqueeze(1)
        faia = pred(ra, stress)
        faib = pred(rb, stress)
        dfaidra = grad(faia, ra, retain_graph=True, create_graph=True)[0]
        dfaidrb = grad(faib, rb, retain_graph=True, create_graph=True)[0]
        J_ex = 2*np.pi*(-Ura*dfaidra + Urb*dfaidrb) #

        loss = J_dom - J_ex
        optim.zero_grad()
        loss.backward(retain_graph=True)
        return loss
    optim.step(closure)
    step_scheduler
```

Listing 18: The testing process of DCEM-O

Interestingly, when $\nu = 0.31$—close to the lower extrapolation boundary—the computation time for DCEM-O is slightly longer than for other values. However, at $\nu = 0.48$, which is near the upper extrapolation boundary, the training time is shorter. This non-monotonic behavior could be attributed to the highly non-convex nature of neural network optimization, suggesting that convergence patterns may not always be deterministic.

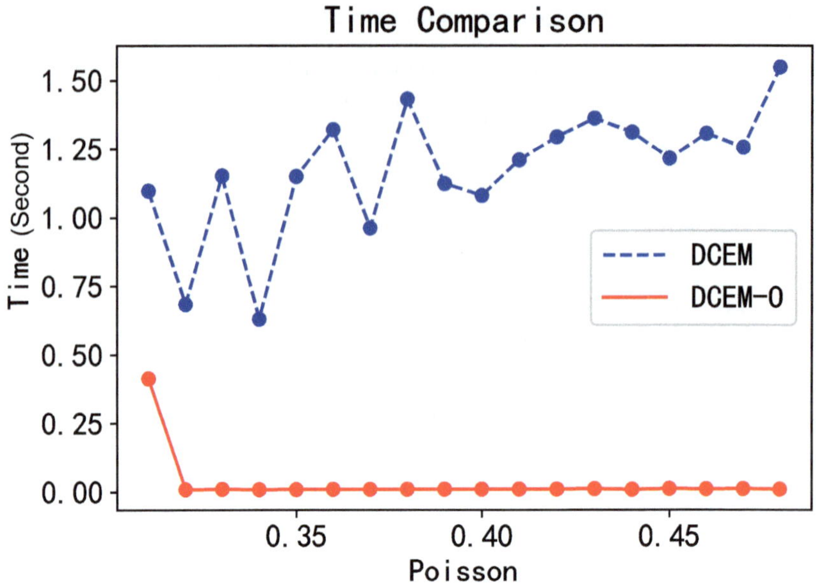

Fig. 6.7 Comparison of DCEM and DCEM-O across different Poisson's ratios: the training time is recorded when the relative $\mathcal{L}_2$ error reaches 1%

6.3 Application of FNO in the Finite Element Method

In this section, we demonstrate how operator learning—particularly the Fourier Neural Operator (FNO)—can be integrated with the finite element method (FEM).

We present a 3D solid mechanics homogenization example. In traditional methods, computing the effective elastic tensor involves solving 18 displacement fields using FEM under six distinct loading conditions (three directions—x, y, and z—for each type of loading). For every new geometry or material configuration (e.g., different Poisson's ratios or Young's moduli), the FEM procedure must be repeated. As the material becomes more heterogeneous and complex, FEM-based computations become increasingly time-consuming, making displacement field prediction a significant bottleneck in homogenization efficiency.

To address this, we adopt the FNO as the backbone of the PreFine-Homo framework [19], as illustrated in the pretraining phase of Fig. 6.8. FNO has gained widespread attention in the AI for PDEs community, especially in high-impact applications such as weather forecasting [20, 21]. Given its success in modeling complex spatiotemporal mappings in fluid mechanics, we argue that it is equally well-suited for solid mechanics problems, which also rely on function-to-function mappings in theory.

The core idea of the PreFine-Homo framework is simple: we first generate a large dataset by varying Poisson's ratios and geometric configurations. The input consists of material matrices, while the output consists of the corresponding 18 displacement

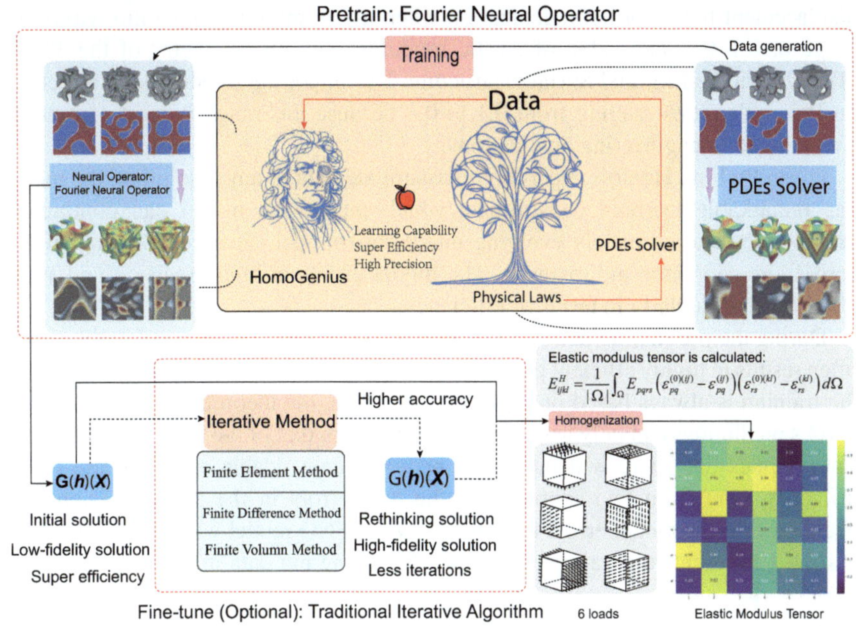

Fig. 6.8 PreFine-Homo consists of two key steps: **pretraining** and **fine-tuning**. In the pretraining phase, a large dataset with varying materials and geometries is generated using traditional numerical solvers. The Fourier Neural Operator (FNO) is then trained on this dataset. In the fine-tuning phase, the FNO provides an initial solution, which serves as the starting point for further refinement using traditional iterative solvers. This two-step framework enables rapid prediction of displacement fields for unseen geometries and materials. Fine-tuning is optional and applied as needed. The effective elastic tensor is subsequently computed using a homogenization formula

fields under six loading conditions (three directions for each). The FNO is trained to learn the mapping from the material matrix (a 3D tensor of resolution $128 \times 128 \times 128$ or $64 \times 64 \times 64$, filled with zeros and Poisson's ratio values) to the corresponding displacement fields. As we focus on linear elasticity, Young's modulus is set to 1 during training, allowing us to obtain the effective elastic tensor by linearly scaling with the actual modulus.

Using the dataset obtained from the traditional homogenization process, we began training our PreFine-Homo. The training time for a resolution of 128 is about 5 hours per 100 epochs (3D FNO with mode 18 on one V100: 32G), and the training time for a resolution of 64 is about 3 hours per 100 epochs (3D FNO with mode 32 on one V100: 32G). The mode of FNO and the amount of data can be configured according to hardware conditions. The model takes a material matrix with four channels as input—specifically, one for the material's Poisson's ratio and three for the spatial coordinates (x, y, and z)—and produces an 18-channel displacement field as output. To address the varying magnitudes across different displacement fields, we implemented normalization techniques to ensure consistency across the displacement channels. Our model is capable of rapidly predicting the 18 necessary displacement fields for homogenization when given any Poisson's ratio within the range of 0.1–0.4, any value of elastic modulus, and any geometry of the TPMS (Triply Periodic Minimal Surface) structure. It is important to note that we focused on Poisson's ratios ranging from 0.1 to 0.4 because this range is most commonly encountered in engineering applications.

Given PreFine-Homo's capability to assimilate data of any resolution, the model has remarkable learning capabilities, which means the model can continually incorporate data, ultimately evolving into a supermodel capable of homogenizing across any geometry and material. The model excels in high precision and super efficiency, particularly in terms of efficiency.

During the pretraining phase, PreFine-Homo is trained on a large dataset, which often results in highly efficient performance during the testing phase. However, since the training is always based on high-precision data, the theoretical accuracy of the FNO's predictions can only asymptotically approach that of the high-precision data. This implies that there will inevitably be some unavoidable errors. These errors primarily stem from two sources: (1) the data errors in the high-precision data itself, and (2) the model errors introduced by the FNO model when fitting the data. Listing 19 presents the training code for FNO using big data in the PreFine-Homo framework.

To address this, we introduce the fine-tuning phase in PreFine-Homo, as shown in the fine-tuning phase of Fig. 6.8. The core idea of the fine-tuning phase is to use the FNO's prediction as the initial value for an iterative algorithm. Since the operator learned during the pretraining phase has been exposed to a vast amount of data, the model can provide a highly accurate initial solution. This initial solution is much closer to the true value compared to a random prediction, thereby significantly reducing the number of iterations required. Note that, theoretically, the fine-tuned solution can achieve arbitrary precision, provided that the threshold in the iterative algorithm is set sufficiently small. Listing 20 presents the fine-tuning code for FNO using big data in the PreFine-Homo framework.

The iterative algorithm used in the fine-tuning phase is determined based on traditional numerical methods for solving PDEs. In computational mechanics, these traditional numerical methods primarily include the Finite Element Method (FEM), Finite Difference Method (FDM), and Finite Volume Method (FVM). The choice of traditional numerical method determines the matrix equation in the iterative algorithm:

$$\boldsymbol{K}\boldsymbol{X} = \boldsymbol{f}, \tag{6.22}$$

where $\boldsymbol{K}$ is determined by the selected numerical method, meaning that different numerical methods for solving PDEs will result in different $\boldsymbol{K}$. Here, $\boldsymbol{X}$ is the vector to be solved, and $\boldsymbol{f}$ is the known vector.

There are two numerical approaches to solving Eq. (6.22): direct methods and iterative methods. For large-scale problems, iterative methods are often preferred due to their computational complexity. Since we are considering 3D problems with over one million degrees of freedom, iterative methods are more suitable than direct

```
model = FNO3d(modes, modes, modes, width).cuda()
print(count_params(model))
optimizer = torch.optim.Adam(model.parameters(), \
                lr=learning_rate, weight_decay=1e-4)
scheduler = torch.optim.lr_scheduler.StepLR(optimizer,\
                    step_size=step_size, gamma=gamma)
start_time = default_timer()
myloss = LpLoss(size_average=False)
y_normalizer.cuda()
for ep in range(epochs):
    model.train()

    train_mse = 0
    for x, y in train_loader:
        x, y = x.cuda(), y.cuda()
        optimizer.zero_grad()
        out = model(x).reshape(batch_size, s_train, \
                                s_train, s_train, 18)
        mse = F.mse_loss(out.view(batch_size, -1), \
              y.view(batch_size, -1), reduction='mean')
        mse.backward()
        out = y_normalizer.decode(out)
        y = y_normalizer.decode(y)
        loss = myloss(out.view(batch_size,-1), \
                            y.view(batch_size,-1))
        # loss.backward()
        optimizer.step()
    scheduler.step()
```

Listing 19: The demo of the pretraining phase of the PreFine-Homo

```
fno_pred = load(fno_predict_folderpath).FNO_X;
[dis] = homo3D_Alloutput(lx,ly,lz,lambda,mu,data, \
                                            fno_pred, tol, initial_mark);

function [dis] = homo3D_Alloutput(lx,ly,lz,lambda,\
                            mu,voxel, initial, tol, initial_mark)
%%%%%%%%%%%%%%%%%%%%%%%%%%%%%%%%%%%%%%%%%%%%%%%%%%%%%%%%%%%%%%%%%%%%%%%%%%%
% lx      = Unit cell length in x-direction.
% ly      = Unit cell length in y-direction.
% lz      = Unit cell length in z-direction.
% lambda  = Lame's first parameter for solid materials.
% mu      = Lame's second parameter for solid materials.
% voxel   = Material indicator matrix. Used to
%           determine nelx/nely/nelz
% dis         = The 18 dis field under 6 loads
%%%%%%%%%%%%%%%%%%%%%%%%%%%%%%%%%%%%%%%%%%%%%%%%%%%%%%%%%%%%%%%%%%%%%%%%%%%%
%% INITIALIZE
```

Listing 20: The demo of the finetuning phase of the PreFine-Homo methods due to their lower complexity. The 3D elastic problem involves large, sparse systems of linear equations, our proposed PreFine-Homo framework employs the Preconditioned Conjugate Gradient (PCG) method as its iterative solver. The iterative formulation for solving Eq. (6.22) is:

$$\boldsymbol{X}^{(k+1)} = \boldsymbol{\phi}(\boldsymbol{X}^{(k)}; \boldsymbol{K}, \boldsymbol{f}), \tag{6.23}$$

where $\boldsymbol{X}^{(k)}$ is the solution vector at the k-th iteration. The iteration stops when the relative error $\mathcal{L}_{rel}$ falls below a threshold tol:

$$\mathcal{L}_{rel} = \frac{\|\boldsymbol{K}\boldsymbol{X} - \boldsymbol{f}\|}{\|\boldsymbol{f}\|} < tol, \tag{6.24}$$

where $\| \cdot \|$ denotes the L_2-norm.

The core of the fine-tuning phase in PreFine-Homo is to use the pretrained output $\boldsymbol{X}^{fno}$ as the initial solution for Eq. (6.23):

$$\boldsymbol{X}^{(k+1)} = \boldsymbol{\phi}(\boldsymbol{X}^{(k)}; \boldsymbol{K}, \boldsymbol{f}, \boldsymbol{X}^{(0)} = \boldsymbol{X}^{fno}). \tag{6.25}$$

The pretraining and fine-tuning phases of PreFine-Homo are conceptually similar to the pretraining and reinforcement learning stages of large language models [22]. Since the pretraining phase of PreFine-Homo provides a high-quality initial solution, the number of iterations in the fine-tuning phase can be significantly reduced compared to traditional methods that use random initial vectors. A reduction in the number of iterations means a decrease in computational time. Theoretically, the fine-tuned solution can achieve arbitrary precision, provided that the convergence threshold in the iterative algorithm is set sufficiently small. It is worth noting that the fine-tuning phase is optional and can be applied based on specific requirements. To determine whether the fine-tuning phase is necessary, we propose evaluating the initial solution provided by the pretraining phase of PreFine-Homo by substituting it into Eq. (6.24) and examining the value of $\mathcal{L}_{rel} = \|\boldsymbol{K}\boldsymbol{X}^{fno} - \boldsymbol{f}\|/\|\boldsymbol{f}\|$. Since $\mathcal{L}_{rel}$ reflects the accuracy of the initial solution, if $\mathcal{L}_{rel} < tol_{fine}$, the fine-tuning phase is deemed unnecessary, where tol_{fine} serves as the threshold for deciding whether to proceed with the fine-tuning phase.

If a certain level of error is acceptable, the pretrained model of PreFine-Homo can be used directly. However, if high-precision solutions are desired, the fine-tuning phase can be employed at the cost of additional computational resources.

PreFine-Homo combines data-driven and physics-based approaches, embodying a dual-thinking paradigm. The framework of PreFine-Homo is highly flexible, capable of delivering homogenization results regardless of the availability of data. Moreover, as the amount of data increases, the accuracy of the operator learned during the pretraining phase of PreFine-Homo improves, leading to a reduction in the number of iterations required during the fine-tuning phase. Therefore, PreFine-Homo represents a novel homogenization computational framework that continuously learns and evolves over time.

We evaluate the generalization ability of PreFine-Homo for different geometries. To assess the performance of PreFine-Homo across various geometries, we employed a range of periodic structures known as Triply Periodic Minimal Surfaces (TPMS). TPMS represents a classical type of periodic structure in artificial cellular materials, characterized by its topology-driven properties.

We generated 1800 TPMS unit cells with resolution 128*128*128 (The resolution of the training and testing sets are both 128) for studying the proposed homogenization method. These 1800 unit cells are divided into 3 groups ("Sheet-networks": Schoen Gyroid, Schwarz Diamond, and Fischer Kosh S). For each TPMS category, we acquired 600 different geometric structure data sets with varying volume fractions. Specifically, the volume fraction of datasets varies from 0.26 to 0.66. We shuffled all the data to divide it into training and testing sets, ensuring that the test set does not appear in the training set. These datasets were then processed through a traditional finite element program for homogenization to calculate the corresponding displacement fields and the effective elastic tensor.

The input data were the Poisson's ratios, meaning the 3D matrix contained zeros where there was no material, and the corresponding Poisson's ratio values at material points. The FNO model's mode is set to 12, width to 32, with a total of 28,322,738 parameters, and a learning rate of 0.001. These parameters can be adjusted according to the hardware capabilities. Our dataset consists of 150 samples each of $128 * 128 * 128$ resolution data from three types of "Sheet-networks": Schoen Gyroid, Schwarz Diamond, and Fischer Kosh S. The input has 4 channels, representing Poisson's ratio, and the x, y, and z coordinates. The output consists of 18 displacement fields, the 3 displacement fields under 6 different loading conditions. Currently, we use a GPU with 32GB of memory, so our input data is limited to 450 samples of 128-resolution data. With multi-GPU training, the dataset can be expanded. Despite using a single GPU, our results are already outstanding. The training set consists of 400 samples, while the test set contains 50 samples. Please note that we have combined all types of TPMS in training and test sets to demonstrate the powerful fitting capability of PreFine-Homo.

We first demonstrate the capabilities of PreFine-Homo in the pretraining phase. Figure 6.9a shows geometric data for Schoen Gyroid of "sheet-networks". We kept the Poisson's ratio constant to focus on the PreFine-Homo model's generalization ability across different geometries. The PreFine-Homo in the pretraining phase generated the 18 needed displacement fields for the homogenization process (three displacement fields for each of the six types of loading).

Due to the extensive amount of model output, we showcased a comparison between PreFine-Homo and the reference solution from FEM under the first type (x-direction tension) of loading. Figure 6.9a, b, and c respectively shows x, y, and z direction displacement field in the test set. It is evident that PreFine-Homo in the pretraining phase closely matches the FEM reference solution.

To highlight the performance of PreFine-Homo in the pretraining phase on TPMS geometries, Fig. 6.10 shows the comparisons in the test set with two other common types of TPMS: Schwarz Diamond and Fischer-Koch S of "sheet-networks". PreFine-Homo closely matches the FEM solutions on the test set across

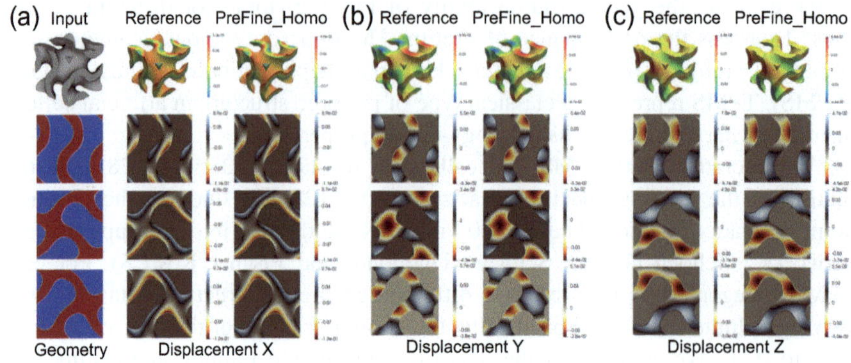

Fig. 6.9 Prediction results of the proposed PreFine-Homo model in the pretraining phase across different geometries (Resolution 128) for Schoen Gyroid of "sheet-networks" in the test set under uniaxial tension in the x-direction. (**a**) The left column shows the geometric input for PreFine-Homo, and the right two columns show the FEM reference solution and the prediction by PreFine-Homo for the x-direction displacement field, y-direction displacement field (**b**), z-direction displacement field (**c**). The second, third, and fourth rows of (**a**, **b**, **c**) show the results for different 2D cross-sections

all three types of TPMS, with an average relative error of the effective elastic tensor of only about 1.58%.

Next, we present the error evolution of FNO in predicting the displacement fields during the homogenization process. Figure 6.11 shows the evolution trends of the training and testing errors. Overall, we found that the testing errors converge to approximately 3% to 5%.

Above, we demonstrated the accuracy and error evolution. Now, we focus on the efficiency. The greatest advantage of PreFine-Homo is its efficiency, as it can quickly predict the displacement fields needed for the homogenization process.

As shown in Table 6.1, the efficiency gains of PreFine-Homo in the pretraining phase are significantly higher compared to the Finite Element Method (FEM). Since FEM relies on iterative algorithms, its computational time depends heavily on the convergence threshold (tol). For instance, when $tol = 10^{-10}$, the average computational time for FEM is approximately 4000 seconds. In contrast, PreFine-Homo in the pretraining phase requires only about 0.43 seconds, demonstrating a remarkable improvement in computational efficiency for displacement field prediction.

Moreover, as illustrated in Fig. 6.11, PreFine-Homo maintains high accuracy, with testing errors converging to approximately 3% to 5%. This combination of efficiency and precision makes PreFine-Homo a powerful tool for large-scale homogenization tasks, significantly outperforming traditional methods in both speed and reliability.

The key to improving homogenization efficiency is to accelerate the prediction of the displacement field required in homogenization. The core reason that PreFine-Homo is faster than traditional algorithms lies in the utilization of operator learning

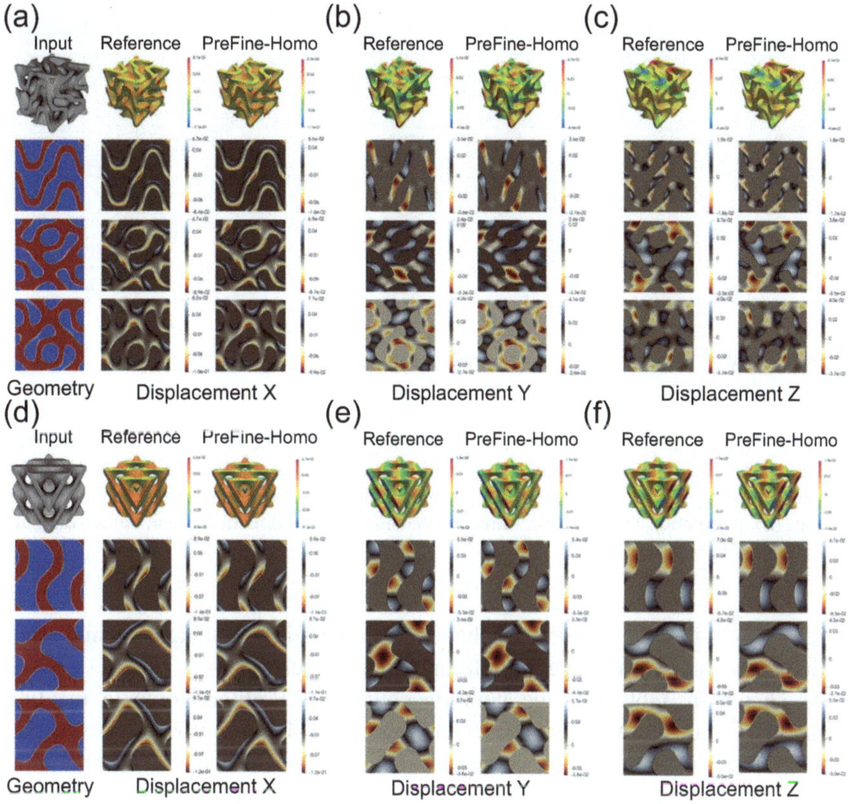

Fig. 6.10 Prediction results of the proposed PreFine-Homo model in the pretraining phase across different geometries (Resolution 128) for Schwarz Diamond (**a**, **b**, **c**) and Fischer-Koch S (**d**, **e**, **f**) of both "sheet-networks" in the test set under uniaxial tension in the x-direction

to boost the efficiency of displacement field prediction, thereby enhancing the overall homogenization efficiency.

Next, we demonstrate the capabilities of PreFine-Homo in the fine-tuning phase. The core idea of PreFine-Homo in the fine-tuning phase is to use the predictions from the pretraining phase as the initial solution for the iterative algorithm. It is worth noting that the fine-tuning phase of PreFine-Homo is optional. If higher precision is not required and efficiency is the primary concern, one can skip the fine-tuning phase and rely solely on the pretraining phase of PreFine-Homo.

A key advantage of PreFine-Homo in the fine-tuning phase lies in its ability to significantly reduce the number of iterations required for convergence compared to traditional PDE solvers like the Finite Element Method (FEM). This reduction is achieved by leveraging the high-quality initial solution provided by the pretraining phase of PreFine-Homo. By starting with an initial solution that is already close to the true solution, PreFine-Homo in the fine-tuning phase can achieve convergence

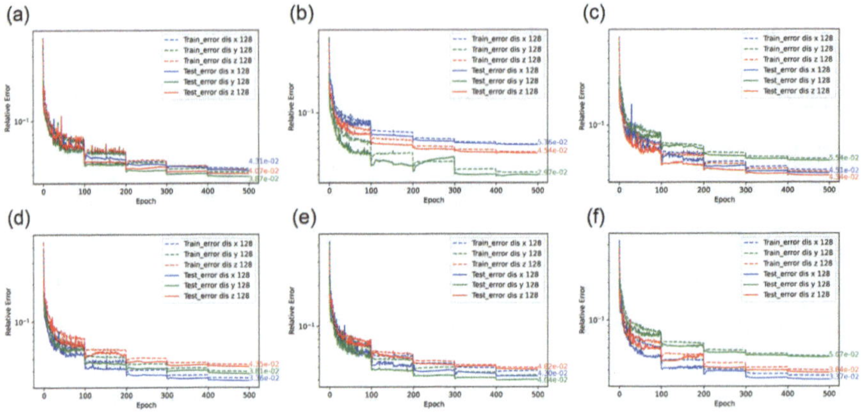

Fig. 6.11 Training and testing errors of the pretraining phase of the PreFine-Homo for the three displacement fields under six loading conditions in different geometries (Resolution 128). The numbers in the figures represent the final converged values of the testing errors: (**a**) x-direction tension, (**b**) y-direction tension, (**c**) z-direction tension, (**d**) xy-direction shear, (**e**) xz-direction shear, (**f**) yz-direction shear

with far fewer iterations, thereby enhancing computational efficiency while maintaining high accuracy as same as traditional PDE solvers.

As shown in Table 6.1, the iteration counts for PreFine-Homo in the fine-tuning phase are substantially lower than those of FEM across various convergence thresholds (tol). The iteration counts are statistically averaged across the entire test set, which includes 50 data points from three types of TPMS structures. For example, at $tol = 10^{-5}$, FEM requires an average of 238.1, 203.6, and 193.8 iterations for Schoen Gyroid (SG), Schwarz Diamond (SD), and Fischer-Koch S (FKS) geometries, respectively. In contrast, PreFine-Homo_Fine requires only 84.28, 70.11, and 65.16 iterations for the above geometries, representing a reduction of approximately 65%, 66%, and 66%, respectively.

This trend holds even for stricter convergence thresholds. At $tol = 10^{-10}$, FEM requires 523.6, 510.7, and 487.2 iterations for SG, SD, and FKS, respectively, while PreFine-Homo_Fine requires 500.2, 488.2, and 473.4 iterations. Although the reduction in iteration counts is less pronounced at higher precision levels as shown in Fig. 6.12, this is because the benefits of a good initial solution diminish as tol decreases. The increased number of iterations required for convergence at higher precision levels tends to overshadow the early-stage benefits provided by the initial solution. Nevertheless, PreFine-Homo still demonstrates a clear advantage, particularly when combined with its significantly faster pretraining phase.

The reduced iteration counts in the fine-tuning phase of PreFine-Homo directly translate to improved computational efficiency, making it a powerful tool for large-scale 3D elasticity problems. It is important to emphasize that the fine-tuning phase of PreFine-Homo ensures accuracy comparable to that of traditional high-precision PDE solvers. As a result, the error in the effective elastic tensor can approach

Table 6.1 The efficiency of PreFine-Homo on an average over all different geometry test cases (a single test case) with a resolution of 128. *tol* is the convergence threshold for the iterative algorithm. FEM represents the average time and average number of iterations per load case for a traditional PDEs solver. PreFine-Homo_Pre denotes the average time for testing a single case in the pretraining phase of PreFine-Homo. PreFine-Homo_Fine represents the average time and average number of iterations per load case for testing a single case in the fine-tuning phase of PreFine-Homo. SG, SD, and FKS refer to Schoen Gyroid, Schwarz Diamond, and Fischer-Koch S of "sheet-networks," respectively

	$tol = 10^{-5}$			$tol = 10^{-10}$		
	SG	SD	FKS	SG	SD	FKS
FEM:sec/iters	1809/238.1	1726/203.6	1592/193.8	3687/523.6	4757/510.7	3558/487.2
PreFine-Homo_Pre:sec	0.42365	0.43983	0.44476	0.42365	0.43983	0.44476
PreFine-Homo_Fine:sec/iters	**867.4/84.28**	**736.3/70.11**	**681.1/65.16**	**3504/500.2**	**4514/488.2**	**3276/473.4**

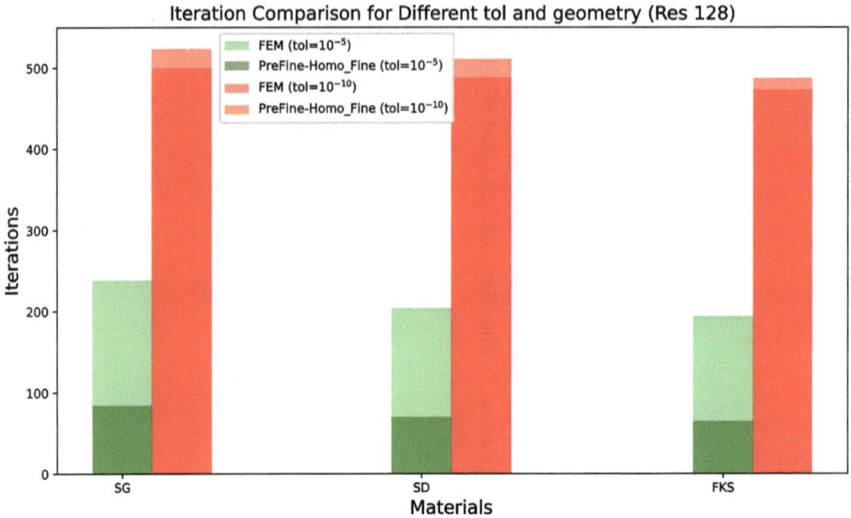

Fig. 6.12 Comparison of the number of iterations required between the traditional PDE solver FEM and the fine-tuning phase of PreFine-Homo across different tolerances (tol) and all different geometry test cases (a single test case) in the iterative algorithm

zero. Therefore, we do not present the homogenization results for the effective elastic tensor here, as its accuracy is entirely determined by the precision of the displacement field.

PreFine-Homo uses the prediction provided by FNO as the initial guess for the Preconditioned Conjugate Gradient (PCG) solver, which shares similarities with traditional preconditioned iterative algorithms but is not exactly the same. Conventional preconditioning aims to approximate the inverse of the stiffness matrix, thereby decreasing the condition number of the original linear system and accelerating convergence.

In PreFine-Homo, since the linear system is highly sparse and positive definite, the PCG method is adopted as the main iterative solver. Because PCG inherently includes a preconditioning mechanism, we also evaluate the performance of PreFine-Homo without any preconditioning, for which the iterative solver is switched to Conjugate Gradient (CG). Figure 6.13 compares the number of iterations between PCG and CG solvers when using FNO-predicted initial guesses. It can be observed that, for both PCG and CG, the number of iterations decreases significantly when FNO is used as the initial guess, especially at higher resolutions, since a higher resolution implies that the initial solution provided by FNO is closer to the true solution.

Furthermore, regardless of whether the FNO prediction is employed as the initial guess, the convergence rate is primarily determined by the iterative algorithm itself. The PCG solver exhibits a faster convergence rate than the CG solver. In essence, the iterative algorithm governs the convergence rate, whereas the initial guess

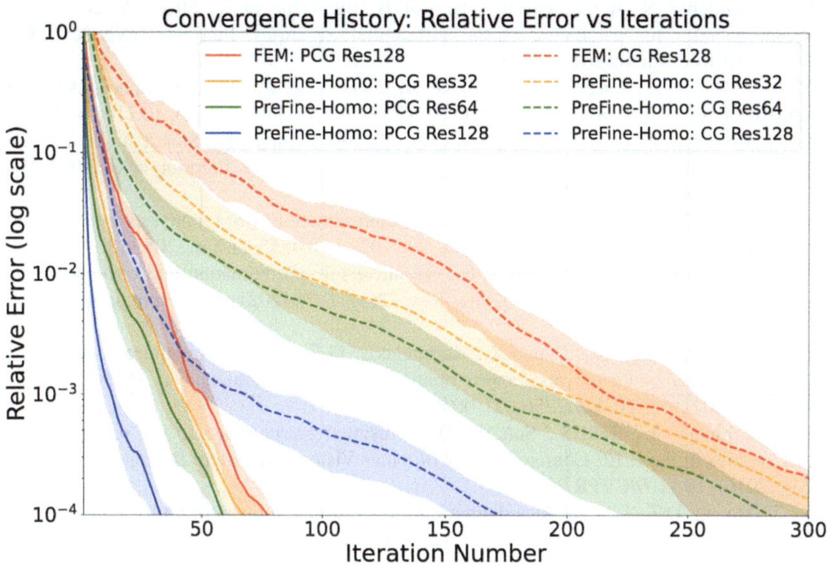

Fig. 6.13 Comparison of iteration numbers under different iterative solvers. The tolerance is set to 10^{-4}, and the solvers are either PCG or CG. For each solver, experiments are conducted both with and without FNO-based initial guesses at different resolutions. "PreFine-Homo" denotes the cases with FNO-based initial guesses, whereas "FEM" corresponds to those without FNO initialization. "Res 32", "Res 64", and "Res 128" represent the resolutions of the training datasets used for the FNO model. Since six independent loading cases are applied, each method yields six iteration counts, and the curves in the figure represent their ranges

determines the starting point of the iteration. It is worth noting that all test cases in our experiments were conducted at a resolution of 128.

The relationship between the PreFine-Homo and preconditioners can be intuitively understood as a mountain-climbing analogy: both aim to reach the top of the mountain (i.e., the converged solution). A preconditioner changes the shape of the mountain, making the slope smoother and the path easier to climb, while our method providing a better initial guess corresponds to starting from a higher point on the mountain, thus reducing the climbing distance needed to reach the summit.

References

1. Kovachki, N.B., Li, Z., Liu, B., Azizzadenesheli, K., Bhattacharya, K., Stuart, A.M., et al. Neural operator: learning maps between function spaces with applications to PDEs. J. Mach. Learn. Res. **24**(89), 1–97 (2023). http://jmlr.org/papers/v24/21-1524.html
2. Chen, T., Chen, H.: Universal approximation to nonlinear operators by neural networks with arbitrary activation functions and its application to dynamical systems. IEEE Trans. Neural Netw. **6**(4), 911–917 (1995). 10.1109/72.392253

3. Li, Z., Kovachki, N., Azizzadenesheli, K., Liu, B., Bhattacharya, K., Stuart, A., et al. Fourier neural operator for parametric partial differential equations. Preprint. arXiv:2010.08895 (2020). 10.48550/arXiv.2010.08895
4. Yang, L., Zhang, Z., Song, Y., Hong, S., Xu, R., Zhao, Y., et al.: Diffusion models: A comprehensive survey of methods and applications. ACM Comput. Surv. (2022). 10.1145/3626235
5. Li, Z., Kovachki, N., Azizzadenesheli, K., Liu, B., Bhattacharya, K., Stuart, A., et al.: Neural operator: Graph kernel network for partial differential equations. Preprint. arXiv:2003.03485 (2020). 10.48550/arXiv.2003.03485.
6. De Ryck, T., Mishra, S.: Generic bounds on the approximation error for physics-informed (and) operator learning. Preprint. arXiv:2205.11393 (2022). 10.48550/arXiv.2205.11393
7. Kovachki, N., Lanthaler, S., Mishra, S.: On universal approximation and error bounds for Fourier neural operators. J. Mach. Learn. Res. **22**(1), 1–76 (2021). http://jmlr.org/papers/v22/21-0806.html
8. Lu, L., Jin, P., Pang, G., Zhang, Z., Karniadakis, G.E.: Learning nonlinear operators via DeepONet based on the universal approximation theorem of operators. Nat. Mach. Intell. **3**(3), 218–229 (2021). 10.1038/s42256-021-00302-5
9. He, K., Zhang, X., Ren, S., Sun, J.: Deep residual learning for image recognition. In: Proceedings of the IEEE Conference on Computer Vision and Pattern Recognition, pp. 770–778 (2016). 10.1109/CVPR.2016.90
10. Li, Z., Huang, D.Z., Liu, B., Anandkumar, A.: Fourier neural operator with learned deformations for PDEs on general geometries. J. Mach. Learn. Res. **24**(388), 1–26 (2023). http://jmlr.org/papers/v24/23-0064.html
11. Lu, L., Meng, X., Cai, S., Mao, Z., Goswami, S., Zhang, Z., et al.: A comprehensive and fair comparison of two neural operators (with practical extensions) based on FAIR data. Comput. Methods Appl. Mech. Eng. **393**, 114778 (2022). 10.1016/j.cma.2022.114778
12. Stumpf, H.: The principle of complementary energy in nonlinear plate theory. J. Elast. **6**(1), 95–104 (1976). 10.1007/BF00135181
13. Wang, Y., Sun, J., Rabczuk, T., Liu, Y.: DCEM: A deep complementary energy method for linear elasticity. Int. J. Numer. Methods Eng. **125**(24), e7585 (2024). 10.1002/nme.7585
14. Goswami, S., Yin, M., Yu, Y., Karniadakis, G.E.: A physics-informed variational DeepONet for predicting crack path in quasi-brittle materials. Comput. Methods Appl. Mech. Eng. **391**, 114587 (2022). 10.1016/j.cma.2022.114587
15. Li, Z., Zheng, H., Kovachki, N., Jin, D., Chen, H., Liu, B., et al.: Physics-informed neural operator for learning partial differential equations. ACM/JMS J. Data Sci. **1**(3), 1–27 (2024). 10.1145/3648506
16. Wang, S., Wang, H., Perdikaris, P.: Learning the solution operator of parametric partial differential equations with physics-informed DeepONets. Sci. Adv. **7**(40), eabi8605 (2021). 10.1126/sciadv.abi8605
17. Vullo, V.: Circular Cylinders and Pressure Vessels. Stress Analysis and Design. Springer, Berlin/Heidelberg (2014). 10.1007/978-3-319-00690-1
18. Cen, S., Fu, X.-R., Zhou, M.-J.: 8- and 12-node plane hybrid stress-function elements immune to severely distorted mesh containing elements with concave shapes. Comput. Methods Appl. Mech. Eng. **200**(29–32), 2321–2336 (2011). 10.1016/j.cma.2011.04.014
19. Wang, Y., Li, X., Yan, Z., Du, Y., Bai, J., Liu, B., et al.: A pretraining-finetuning computational framework for material homogenization. Preprint. arXiv:2404.07943 (2024). 10.48550/arXiv.2404.07943
20. Bi, K., Xie, L., Zhang, H., Chen, X., Gu, X., Tian, Q.: Accurate medium-range global weather forecasting with 3D neural networks. Nature **619**(7970), 533–538 (2023). 10.1038/s41586-023-06185-3
21. Pathak, J., Subramanian, S., Harrington, P., Raja, S., Chattopadhyay, A., Mardani, M., et al.: FourCastNet: A global data-driven high-resolution weather model using adaptive Fourier Neural Operators. Preprint. arXiv:2202.11214 (2022). 10.48550/arXiv.2202.11214
22. Guo, D., Yang, D., Zhang, H., Song, J., Wang, P., Zhu, Q., et al.: Deepseek-R1 incentivizes reasoning in LLMs through reinforcement learning. Nature **645**(8081), 633–638 (2025). 10.1038/s41586-025-09422-z

Chapter 7
Conclusions and Future Directions

7.1 Summary of Key Developments

The focus of this book has been on scientific machine learning for engineering applications, particularly for the solution of partial differential equations, surrogate modeling, optimization and inverse analysis. Solving partial differential equations with machine learning methods can be traced back to the 1990s and the seminal manuscript by Lagaris [1], though its drastically growing interest has started only a few years ago, driven by vast improvements in GPU computing and machine learning approaches such as deep neural networks [2] and associated computational tools such as TensorFlow and PyTorch.

Throughout this book, we have explored how machine learning is revolutionizing computational mechanics through several paradigm shifts. Physics-informed neural networks embed governing equations directly into the learning process, enabling solutions with sparse or no data. Reduced-order modeling enhanced by autoencoders and recurrent networks achieves computational speedups of several orders of magnitude while maintaining fidelity. Digital twins integrate real-time sensor data with predictive models to enable continuous monitoring, predictive maintenance, and optimal control of complex systems. These advances collectively demonstrate that machine learning is not merely a tool for data analysis but a fundamental methodology for scientific computing.

7.2 Machine Learning Interatomic Potentials

There are more interesting areas to explore including machine learning interatomic potentials (MLIPs) or deep potential molecular dynamics (DPMD) [3] that are of major importance in computational materials design. The design of new materials

T. Rabczuk et al., *Scientific Machine Learning with Engineering Applications*,
Studies in Systems, Decision and Control 667,
https://doi.org/10.1007/978-3-032-20307-6_7

and systems involves the optimization of the system's topology as well as the optimization of the materials used for the system. For example, the macroscopic properties of a material are the main criteria for its choice for a particular device. However, the origins of macroscopic material properties reside in the properties and interactions taking place on finer scales. For the design of new materials, it is therefore essential to understand and quantify the links between phenomena on the fine scales and their macroscopic effects. As experimental manufacturing and testing is time-consuming, expensive and sometimes unfeasible, computational multiscale methods are needed.

Approaches such as MLIPs or DPMD promise to bridge the gap between engineering and materials science as they allow connecting the finest quantum mechanical length scales up to the macroscopic length scale through hierarchical multiscale modeling. For specific applications, these approaches reach the accuracy of DFT (density functional theory) simulations at a fraction of their cost and therefore have the potential to be a real game changer. The idea is conceptually straightforward: MLIPs employ a function that can approximate the environment of an atom in a system, which can be used to interpolate the energy, force and stresses. This requires training data obtained through DFT simulations, mostly ab-initio molecular dynamics (AIMD) simulations. Then, the function is fitted to subsampled AIMD data and one can test the accuracy using the full dataset.

Their prospects are:

1. A highly efficient evaluation of key physical (mechanical, thermal, thermodynamics) properties: For instance, all the properties related to phonon dispersion can be extracted from MLIP simulations.
2. Accurate analysis of the interaction between different building blocks, for example the cohesive zone properties or thermal resistance between different materials.
3. Drastic acceleration of the calculations. Together with other ML or optimization algorithms this enables the route for faster finding and designing new materials.
4. Automatization of the design process. Finally, less input data will be required from the user, and materials discovery can be guided by automated active learning frameworks.

7.3 Neural Operators and Operator Learning

Particularly promising is the direction of neural operators, which can not only approximate functions but operators of the solution, potentially accelerating the solution of problem classes that commonly require thousands or more simulations, as in the design of new materials such as composites or specific metamaterials where the microstructure or geometry is key. Fourier Neural Operators (FNO), DeepONet, and related architectures learn mappings between infinite-dimensional function spaces, enabling them to generalize across different discretizations, domains, and boundary conditions [4, 5].

Recent developments have demonstrated that neural operators can achieve performance comparable to traditional numerical methods in computational fluid dynamics, weather forecasting, and material modeling while being 4–5 orders of magnitude faster [6]. Universal Physics Transformers extend this capability by providing an efficient unified learning paradigm for a wide range of spatiotemporal problems [7]. The PROSE framework represents a significant advancement as the first multimodal scientific foundation model capable of zero-shot prediction, generating both numerical solutions and symbolic mathematical equations from diverse inputs [8].

The integration of transformer architectures with operator learning has opened new possibilities for handling complex geometries and multi-physics problems. Physics-aware attention mechanisms enable adaptive domain segmentation and efficient representation of solution features across scales. As these methods mature, they promise to transform how we approach parametric studies, uncertainty quantification, and inverse design problems that require evaluating models across vast parameter spaces.

7.4 Foundation Models for Scientific Computing

Foundation models represent an emerging paradigm that could fundamentally transform scientific computing and engineering design. These large-scale pre-trained models, analogous to GPT and similar architectures in natural language processing, are being adapted for scientific domains to handle diverse tasks including literature review, information extraction, knowledge graph construction, and scientific hypothesis generation [9]. There is also interest in developing models that can act as "AI co-scientists" [10], which have had some recent successes in drug-repurposing [11] and discovering the mechanism of gene transfer in bacteria [12]. Unlike task-specific models, foundation models are trained on massive multi-modal datasets spanning multiple scientific domains, enabling them to transfer knowledge across problems and discover cross-domain patterns.

Scientific foundation models (SciFMs) offer several transformative capabilities. They can integrate information from heterogeneous sources including simulation data, experimental measurements, scientific literature, and domain expertise. Multi-modal architectures enable these models to process and generate both numerical predictions and symbolic expressions, bridging the gap between data-driven learning and symbolic reasoning. The ability to perform zero-shot and few-shot learning allows SciFMs to tackle new problems with minimal task-specific training data, dramatically reducing the cost of model development for new applications.

However, significant challenges remain in developing effective SciFMs for engineering and physical sciences. The requirement for massive computational resources for training raises questions about accessibility, environmental sustainability, and the ultimate effectiveness of these models. Ensuring physical consistency and maintaining accuracy require careful architectural design and training strategies. Uncertainty quantification becomes more complex with large-scale models, and establishing trust in model predictions for safety-critical applications demands rigorous validation and interpretability. Despite these challenges, the potential of foundation models to accelerate scientific discovery and enable new modes of human-AI collaboration makes them a critical research frontier.

7.5 Critical Challenges and Open Problems

Despite remarkable progress, several fundamental challenges must be addressed to realize the full potential of machine learning in computational mechanics and scientific computing. Understanding and overcoming these limitations is essential for transitioning from proof-of-concept demonstrations to robust engineering tools.

7.5.1 Optimization and Trainability

Training physics-informed neural networks and related architectures remains challenging, particularly for stiff equations, multi-scale problems, and high-frequency phenomena. The optimization landscape is often highly non-convex with numerous local minima, and training can be sensitive to initialization, learning rate schedules, and loss term weighting [13]. Spectral bias causes neural networks to preferentially learn low-frequency components, hindering convergence for problems with sharp gradients or discontinuities. Adaptive loss balancing, curriculum learning, and novel network architectures show promise but require further development for broad applicability.

The nonlinear regime poses particular difficulties where neural tangent kernel theory does not apply and training dynamics are poorly understood. Developing optimization methods specifically tailored for scientific machine learning, including better understanding of the loss landscape geometry and effective preconditioning strategies, represents an important research direction. Automatic differentiation through complex simulation codes can be compute and memory-intensive, motivating research into checkpointing strategies and efficient training algorithms.

7.5.2 Uncertainty Quantification and Robustness

Quantifying uncertainty in neural network predictions is essential for engineering applications where decisions carry safety and cost implications. While Bayesian neural networks, ensemble methods, and dropout-based approaches provide uncertainty estimates, their computational cost can be prohibitive for large models [14]. Conformal prediction offers distribution-free uncertainty quantification but requires careful calibration and may produce overly conservative intervals for out-of-distribution inputs.

Robustness to distributional shift and adversarial perturbations is critical for deployment in real-world conditions that differ from training data. Neural networks can exhibit unexpected failures when extrapolating beyond their training domain, and developing reliable indicators of when predictions can be trusted remains an open problem. Combining data-driven learning with physics-based constraints can improve robustness, but the optimal balance and integration strategy depends on problem specifics.

7.5.3 Data Efficiency and Generalization

Many engineering applications involve expensive experiments or simulations, limiting the available training data. Active learning and Bayesian optimization can guide data acquisition to maximize information gain, but these approaches require reliable uncertainty estimates and may struggle with high-dimensional parameter spaces. Transfer learning enables knowledge transfer from data-rich to data-scarce domains, but negative transfer can occur when source and target domains differ significantly.

Incorporating physical knowledge through inductive biases, symmetries, and conservation laws can dramatically improve data efficiency and generalization. However, identifying appropriate structural constraints and implementing them within neural network architectures requires domain expertise and careful design. Self-supervised learning on unlabeled simulation data and physics-guided data augmentation represent promising directions for improving sample efficiency.

7.6 Integration with Traditional Numerical Methods

The future of computational mechanics likely lies not in replacing traditional numerical methods entirely but in developing hybrid approaches that synergistically combine physics-based modeling with data-driven learning. Finite element methods, finite volume methods, and spectral methods embody decades of mathematical and algorithmic development, providing provable convergence, conservation properties, and well-understood error bounds. Machine learning can enhance these methods in several ways.

Neural networks can serve as learned components within larger numerical frameworks, such as approximating subgrid-scale closures in turbulence modeling [15, 16] or providing constitutive models for inelastic materials [17, 18]. Adaptive mesh refinement can be guided by neural network-based error indicators that learn to identify regions requiring higher resolution [19]. Preconditioners for iterative solvers can be learned from previous solutions, accelerating convergence for sequences of related problems [20, 21]. Reduced-order models constructed through autoencoders can provide physics-constrained low-dimensional representations for efficient simulation of parametric systems.

Conversely, traditional numerical methods inform machine learning by providing training data, defining physics-informed loss functions, and suggesting appropriate network architectures. Structure-preserving neural networks that respect symplectic geometry, energy conservation, or other invariants draw inspiration from geometric integrators in numerical analysis. The interplay between discrete and continuous formulations, central to numerical analysis, informs the design of continuous-depth neural networks and neural ordinary differential equations.

Developing software frameworks that seamlessly integrate machine learning with established scientific computing libraries will be essential for adoption by the broader engineering community. Standards for model verification, validation, and uncertainty quantification adapted to hybrid physics-ML methods will build confidence in these approaches for safety-critical applications.

7.7 Future Outlook

Several trends will likely shape the field's evolution over the coming years. Increased integration between machine learning and traditional scientific computing will yield hybrid methods that leverage the complementary strengths of both paradigms. Domain-specific architectures and training strategies tailored to particular problem classes will outperform generic approaches, driving specialization within scientific machine learning. Automated machine learning tools will lower barriers to entry, enabling domain scientists without extensive ML expertise to apply these methods.

Foundation models for science will evolve to handle increasingly complex multi-physics problems, potentially enabling AI assistants that can propose novel designs, generate hypotheses, and even conduct autonomous experiments. However, realizing this vision requires solving challenges in controllability, reliability, and alignment with scientific values.

Real-time and online learning will enable adaptive systems that continuously improve from streaming data, crucial for digital twins and autonomous systems. Edge deployment of neural network models will bring intelligence closer to sensors and actuators, enabling low-latency inference and reducing communication overhead.

Quantum computing, while still in early stages, may eventually accelerate both the training of quantum machine learning models and the simulation of quantum systems using classical ML approaches. The interplay between quantum computing and machine learning represents a fascinating frontier with potential implications for materials discovery and quantum chemistry.

The success of scientific machine learning will be measured not just by algorithmic innovations but by its impact on accelerating scientific discovery, enabling sustainable technologies, and improving quality of life. The journey from proof-of-concept to practical deployment requires sustained collaboration across disciplines, rigorous validation, and commitment to open science principles that make advances broadly available. With these commitments, scientific machine learning can fulfill its promise as a transformative force in computational science and engineering.

References

1. Lagaris, I.E., Likas, A., Fotiadis, D.I.: Artificial neural networks for solving ordinary and partial differential equations. IEEE Trans. Neural Netw. **9**(5), 987–1000 (1998). 10.1109/72.712178
2. LeCun, Y., Bengio, Y., Hinton, G.: Deep learning. Nature **521**(7553), 436–444 (2015). 10.1038/nature14539
3. Zhang, L., Han, J., Wang, H., Car, R., W, E.: Deep potential molecular dynamics: a scalable model with the accuracy of quantum mechanics. Phys. Rev. Lett. **120**(14), 143001 (2018). 10.1103/PhysRevLett.120.143001
4. Li, Z., Kovachki, N., Azizzadenesheli, K., Liu, B., Bhattacharya, K., Stuart, A., et al.: Fourier neural operator for parametric partial differential equations. Preprint. arXiv:2010.08895 (2020). 10.48550/arXiv.2010.08895
5. Lu, L., Jin, P., Pang, G., Zhang, Z., Karniadakis, G.E.: Learning nonlinear operators via DeepONet based on the universal approximation theorem of operators. Nat. Mach. Intell. **3**(3), 218–229 (2021). 10.1038/s42256-021-00302-5
6. Azizzadenesheli, K., Kovachki, N., Li, Z., Liu-Schiaffini, M., Kossaifi, J., Anandkumar, A.: Neural operators for accelerating scientific simulations and design. Nat. Rev. Phys. **6**(5), 320–328 (2024). 10.1038/s42254-024-00712-5
7. Alkin, B., Fürst, A., Schmid, S., Gruber, L., Holzleitner, M., Brandstetter, J.: Universal physics transformers: a framework for efficiently scaling neural operators. Preprint. arXiv:2402.12365 (2024). 10.48550/arXiv.2402.12365
8. Liu, Y., Zhang, Z., Schaeffer, H.: PROSE: Predicting multiple operators and symbolic expressions using multimodal transformers. Neural Netw. **180**, 106707 (2024). 10.1016/j.neunet.2024.106707
9. Subramanian, S., Harrington, P., Keutzer, K., Bhimji, W., Morozov, D., Mahoney, M., et al.: Towards foundation models for scientific machine learning: characterizing scaling and transfer behavior. Preprint. arXiv:2306.00258 (2023). 10.48550/arXiv.2306.00258
10. Gottweis, J., Weng, W.-H., Daryin, A., Tu, T., Palepu, A., Sirkovic, P., et al.: Towards an AI co-scientist. Preprint. arXiv:2502.18864 (2025). 10.48550/arXiv.2502.18864
11. Guan, Y., Cui, L., Inchai, J., Fang, Z., Law, J., Brito, A.A.G., et al.: AI-assisted drug re-purposing for human liver fibrosis. Advanced Science, e08751 (2025). 10.1002/advs.202508751
12. Penadés, J.R., Gottweis, J., He, L., Patkowski, J.B., Daryin, A., Weng, W.-H., et al.: AI mirrors experimental science to uncover a mechanism of gene transfer crucial to bacterial evolution. Cell (2025). 10.1016/j.cell.2025.08.018
13. Farea, A., Yli-Harja, O., Emmert-Streib, F.: Understanding physics-informed neural networks: techniques, applications, trends, and challenges. AI **5**(3), 1534–1557 (2024). 10.3390/ai5030074

14. Abdar, M., Pourpanah, F., Hussain, S., Rezazadegan, D., Liu, L., Ghavamzadeh, M., et al.: A review of uncertainty quantification in deep learning: techniques, applications and challenges. Inf. Fusion **76**, 243–297 (2021). 10.1016/j.inffus.2021.05.008
15. Jakhar, K., Guan, Y., Mojgani, R., Chattopadhyay, A., Hassanzadeh, P.: Learning closed-form equations for subgrid-scale closures from high-fidelity data: promises and challenges. J. Adv. Model. Earth Syst. **16**(7), e2023MS003874 (2024). 10.1029/2023MS003874
16. Buaria, D., Sreenivasan, K.R.: Forecasting small-scale dynamics of fluid turbulence using deep neural networks. Proc. Natl. Acad. Sci. **120**(30), e2305765120 (2023). 10.1073/pnas.2305765120
17. Linka, K., Hillgärtner, M., Abdolazizi, K.P., Aydin, R.C., Itskov, M., Cyron, C.J.: Constitutive artificial neural networks: a fast and general approach to predictive data-driven constitutive modeling by deep learning. J. Comput. Phys. **429**, 110010 (2021). 10.1016/j.jcp.2020.110010
18. Dettmer, W.G., Muttio, E.J., Alhayki, R., Perić, D.: A framework for neural network based constitutive modelling of inelastic materials. Comput. Methods Appl. Mech. Eng. **420**, 116672 (2024). 10.1016/j.cma.2023.116672
19. Chan, C.L., Scholz, F., Takacs, T.: Locally refined quad meshing for linear elasticity problems based on convolutional neural networks. Eng. Comput. **38**(5), 4631–4652 (2022). 10.1007/s00366-022-01677-8
20. Zhang, E., Kahana, A., Kopaničáková, A., Turkel, E., Ranade, R., Pathak, J., et al.: Blending neural operators and relaxation methods in PDE numerical solvers. Nat. Mach. Intell. **6**(11), 1303–1313 (2024). 10.1038/s42256-024-00910-x
21. Kopaničáková, A., Lee, Y., Karniadakis, G.E.: Leveraging operator learning to accelerate convergence of the preconditioned conjugate gradient method. Preprint. arXiv:2508.00101 (2025). 10.48550/arXiv.2508.00101

MIX
Papier aus verantwortungsvollen Quellen
Paper from responsible sources
FSC® C105338

If you have any concerns about our products,
you can contact us on
ProductSafety@springernature.com

In case Publisher is established outside the EU,
the EU authorized representative is:
Springer Nature Customer Service Center GmbH
Europaplatz 3, 69115 Heidelberg, Germany

Printed by Libri Plureos GmbH
in Hamburg, Germany